your brain knows more
than you think

# your brain knows more than you think

## the new frontiers of neuroplasticity

**Niels Birbaumer**
with Jörg Zittlau

**Translated by David Shaw**

SCRIBE
*Melbourne • London*

Scribe Publications
18–20 Edward St, Brunswick, Victoria 3056, Australia
2 John Street, Clerkenwell, London, WC1N 2ES, United Kingdom

Originally published as *Dein Gehirn weiss mehr, als du denkst* in German by
Ullstein in 2014
First published in English by Scribe in 2017
This edition published 2018

Printed and bound in the UK by CPI Group (UK) Ltd, Croydon CR0 4YY

Scribe Publications is committed to the sustainable use of natural resources and
the use of paper products made responsibly from those resources.

9781947534094 (US hardback)
9781925322361 (Australian paperback)
9781911344384 (UK paperback)
9781925548310 (e-book)

CiP records for this title are available from the National Library of Australia
and the British Library.

scribepublications.com
scribepublications.com.au
scribepublications.co.uk

# Contents

## Foreword

# Why We Change

*Panta rhei.* This aphorism was supposedly coined by the ancient Greek philosopher Heraclitus nearly 3,000 years ago. It was his belief that 'everything flows' and nothing remains as it is; the only constant is change. Most of us would agree unreservedly with this idea; after all, we see the world changing every day as we go about our lives — and that's not only true of every*thing*, but of every*one*, too. Children become adults, eloquent professors turn into care-dependent dementia patients, liberal democrats develop into dogmatic ultraconservatives, dutiful wives mutate into man-eating seductresses, loving husbands become violent rapists, high-school dropouts transform into dotcom billionaires, and wallflowers grow into show-stopping stars, only to end up later as alcoholics. When it comes to the life trajectory of any given individual, anything is possible. It's sometimes fascinating, sometimes horrifying, but always very interesting.

Nonetheless, we have a tendency to 'freeze-frame' our fellow human beings in certain situations. We speak of born orators, artists, or thinkers, but we also identify born losers and born criminals. We are unable to believe hardnosed psychopaths can ever become valuable members of society, and demand they be locked up forever. We call for a 'humane' end for bedridden patients with locked-in syndrome or those in a persistent vegetative state; we'd like to put an end to their 'suffering' by switching off the machines their lives depend on. We even doubt that people with attention-deficit disorders, chronic depression, or anxiety will ever be able to 'get their act together' again. The argument we hear repeatedly, even from medical professionals and therapists, is that it is better to put them on medication for the rest of their lives than risk them throwing themselves off a cliff or, even worse, dragging other people into the maelstrom of their dark destiny.

*Panta rhei* — that is the philosophy. It may sound logical and reasonable, but *Homo sapiens* tends very often to assume the precise opposite, and that tendency is particularly strong when we think about matters of the mind and behaviour. It is in precisely that context that we harbour the fatalistic assumption that some people just can't change, and so there can be only one solution for dealing with them: they must somehow be prevented from becoming a burden on society or from hurting themselves and others. By controlling them, punctiliously isolating them, or even locking them away. Just don't let them disturb the peace, that's the main thing.

Friedrich Nietzsche said that, with the conceptual tools available to them, human beings were doomed to fail repeatedly to express the idea of 'becoming'. He argued that this was the case because, although words could to some extent at least be used to capture that which *is*, they are not capable of capturing what *will be*. It may be that this deficiency explains why we repeatedly hear talk of immutable character traits and personal characteristics.

However, if we examine the areas in which it usually occurs, we could just as easily conclude that this attitude towards people stems from fear and laziness. After all, it is easier to lock up psychopaths and throw away the key than to reintegrate them into society and run the risk of them relapsing into their old behaviour patterns. And when locked-in patients are 'switched off', their relatives and friends are spared a great deal of trouble and frustration.

The wife of one of my locked-in patients had to face the fact that the man lying in bed staring at the ceiling and struggling to breathe no longer bore any resemblance to her once-witty and energetic husband. Although he had learned to communicate with his environment via a brain–machine interface (BMI) while in our care, this was not the normality she had expected from him. She told us she would never have agreed to let us attempt to communicate with him if she had known this beforehand. She wrestled with the idea that maybe his life-support measures could still be terminated. Not because she wanted to end her husband's suffering, but because she wanted to end her own.

We may speculate about why we have this tendency

to attribute such immutability to our brain and our behaviour when it suits us. But I am more interested in demonstrating how *wrongheaded* that idea is. This book is about *neuroplasticity* — the virtually limitless capacity of the brain to remould itself. This book explains why neither locked-in patients nor those with depression, addictions, or anxiety disorders and neither hyperactive fidgets nor ruthless psychopaths are frozen forever in their behaviour patterns, immune to any attempt to influence them. It describes how we can change into a compassionate 'character' within a very short time, but it also shows how we can just as quickly go from being a loving family man to an inconceivably cruel mass murderer and then back again to an upstanding citizen. People often like to reinterpret excrescences such as Nazism, and the subsequent attempts to trivialise it, as 'exceptional cases' in world history, when in fact they are nothing other than 'perfectly normal' products of the enormous plasticity of our brains. Humans have been both blessed and cursed with an almost limitless willingness to learn — it is both a boon and our doom.

Whether a better understanding of the brain processes associated with learning really does make them easier to control is something we do not know. But if it weren't for the hope that that is the case, many of the experiments described in this book would never have been carried out. Most of the research activities reported here come from my scientific 'workshop', the Institute of Medical Psychology and Behavioural Neurobiology, in Tübingen, since those are the ones from which I can draw first-hand knowledge and experience.

Looking at the course of history, we see that the increase in our knowledge of the development of human behaviour since the Renaissance and the Enlightenment has not done much to advance self-control or humanisation. Nonetheless, various experiments have shown that 'observational learning', i.e. learning that occurs through observing the behaviour of others, controls one of the most efficient learning processes, and that observational learning is firmly embedded in specialised cells in the cerebrum, which have recently become known by the rather vague term 'mirror neurons'.

There is one problem with this mechanism: our brains 'mindlessly' copy anything that promises success and effect. This is why we must strive to maintain a democratic context to our lives and social living conditions, so that our plastic brains do not seduce us into acts of denunciation, murder, and slaughter as has so often been the case in history. When such behaviours are rewarded — as they usually are in undemocratic, dictatorial systems — the brain will principally conform to those patterns.

Conversely, how can such values as respect, empathy, and tolerance be established? Research into the psychology of learning has shown that this becomes more successful the more we know about the aim and purpose of the behaviour to be learned. This does not mean an ideological or moral superstructure such as is sought not only by democracies, but also by dictatorships. What is meant is rather the conscious creation of a memory of the consequences of a certain behaviour. If we experience a positive effect achieved through compassion and respect — and this is more likely to happen in a democracy than in a dictatorship — those behaviour patterns will become stabilised within us and will recur again and again.

This book is about the brain processes involved in learning, but also about loss of self-control; its individual sections outline which people and situations we must pay particular attention to in that context. Each chapter covers one of the topics and issues in human behaviour I have studied with my collaborators over the decades.

I begin in Chapters 1 and 2 with the self-regulation of the brain, using the example of schizophrenia before moving on to the effect neuroplasticity has on the way we behave, think, and feel. Next, I explore what then remains of our so-called character and whether we can even consider it anything more than a more-or-less-random pattern of habits.

Chapters 3 and 4 consider how it is possible to communicate with completely locked-in, paralysed patients by teaching them how they can use their brains to talk by means of state-of-the-art imaging techniques — and the way in which such interaction repeatedly reveals that these supposedly moribund individuals can indeed experience a very high quality of life. This culminates not only in a plea against the ever louder and hastier demands for such patients to be 'switched off' or to be assisted in dying, but also in a warning about the continued rampant trend for advance healthcare directives, often known as 'living wills'.

Chapter 5 explores the brain's ability to repair itself,

as in the case of epilepsy or following a stroke. The German actor Peer Augustinski, for example, was almost completely paralysed on one side of his body after suffering a stroke — but was able to return to the stage after treatment.

Chapter 6 shows that anxieties can largely be neutralised using training techniques; usually, a course of confrontation therapy is all it takes. Neither highly problematical medicinal drugs nor expensive technical equipment are necessary — although it may sometimes be helpful for the therapist to chain herself to the bed with her client, or to go out scooping up dog poop in the park with him.

That which anxiety patients have too much of is lacking in sufficient quantities in psychopaths: activation of the areas of the brain that cause them to vacillate. Psychopaths also lack compassion and empathy. Which, however, does not stop them from pursuing successful careers. Certainly, many psychopaths can be found in prisons, but they are also often to be found in company boardrooms and in the upper echelons of political parties. On the bright side, Chapter 7 shows that their brains are open for change. This means even psychopaths can learn to empathise and develop a necessary measure of fear.

Even senile dementia — which the pharmaceutical industry and therapists connected with it like to pathologise, for their own financial gain, under the general heading of 'Alzheimer's disease' — can be influenced by teaching patients how to control the processes of their own brains. As described in Chapter 8, music can be a valuable aid to this. Just as it can act generally as a phenomenal force for shaping our plastic brains. Imaging techniques do not necessarily enable us to identify the profession of a brain's owner, but the brain of a musician is almost always recognisable.

Chapter 9 deals with another important issue of our time: attention-deficit hyperactivity disorder (ADHD) in children. This problem can also be controlled using brain-training techniques (such as so-called neurofeedback) without recourse to tranquilisers like Ritalin.

The right training methods can also awaken in any one of us a so-called savant ability — that is, a specialised, exceptional talent — by enabling us to unconsciously control our unconscious mind. That might sound paradoxical at first, but it can work — and Chapter 10 describes how.

Neuroplasticity opens up endless possibilities. But let's not get carried away. Not everything is curable, or possible, and the same is true in the context of the brain. Worse, the brain's enormous plasticity has its dark side: the fact that anything that gratifies, stimulates, delights, rewards, or relaxes our brain — in short, anything that makes us feel good — can become addictive. In principle, we are all potential addicts. Chapter 11 explains the mechanisms in the brain that underlie this, and describes how healthy wishes and desires can metamorphose into relentless compulsion and insatiable craving.

Conversely, understanding the greed mechanisms of the brain can help to release an addict from the cycle of wanting more and ever more, although that may not be enough to attain the state of nirvana, which is the focus of the book's conclusion.

The foundation of each chapter is neuroscientific research carried out by my associates and myself. However, I also refer to the work of other scientists and — no less importantly — to philosophers such as Ludwig Hohl, Friedrich Nietzsche, and, in particular, Arthur Schopenhauer. This is because, throughout my many years of working as a researcher in brain science, I have been struck by the fact that much of what we have discovered in our scientific research was already 'pre-thought' centuries ago by philosophers. Sometimes the novelty of a scientific finding consists in nothing more than providing scientific confirmation of ancient knowledge, and thereby attracting to it the attention it deserves.

I was assisted in connecting the results of brain research and philosophical enquiry, and in formulating and explaining my research work, by the philosopher and science journalist Jörg Zittlau. We both hope that, in reading this book, people will gain a new insight into the amazing organ nestling inside their skull. My late friend, the Italian brain anatomist, who later also worked in Tübingen, Valentino Braitenberg, dubbed that organ a 'thought pump'. This may sound mechanistic, but it wonderfully expresses the fact that, on the one hand, the brain is merely a vehicle of transportation, while, on the other hand, it transports something to the surface that would never have emerged if it weren't for the pump. Or, to express it in Schopenhauer's words: the brain is not all, but without the brain, all is nothing.

Of course, we can stretch the metaphor of the pump as far as we like. But, at this stage, let's be content with one hope: that Jörg and I manage to activate the thought pumps of our readers, and in doing so, that we cause a surprise or two to emerge.

# 1

# What Is Personality?

*from a street gang to university is
just a small step*

The scissors were stuck in his flesh. The guy looked at me, bewildered. We were both taken aback. It's in the nature of an impulsive act that those involved do not really understand it. Not the one who commits it, and certainly not the victim. Now it was too late — this guy now had a pair of scissors sticking out of his foot. He screamed blue murder, and I was in trouble. Of course the teachers called the police.

We have a tendency, as we get older, to furnish our past with a meaningful context. Or, in the words of the Swiss writer Max Frisch, 'Sooner or later, everyone invents a story which they believe to be their life.' Politicians love to tell how they were elected class representative at school and already displayed a talent for debating and public speaking. Writers like to regale us with stories about composing their first poems at the age of ten.

And scientists pride themselves on having prepared a hay infusion and observed the resulting microbes under the microscope when they were still knee-high to a grasshopper, or having pondered centrifugal forces while playing on the swings in the park, a little like Sir Isaac Newton, who is said to have come up with the theory of gravity while sitting under a tree in an orchard, after an apple fell on his head.

I'm afraid I can't offer you stories of any such brilliant, or, at the very least, venerable, childhood accomplishments. Instead, I was a member of a teenage street gang.

We stole, ran riot, broke into cars, and generally thought we were the toughest guys in Vienna. There's nothing special about that, but even that would have been enough to get me into trouble. But we were clever enough not to get caught. So we weren't just tough, we were smart, too. Everyone paid attention to us. In short, we were cool. Or so we thought. In my case, at least, it wasn't the tiniest bit true. I may have been impulsive, but I certainly wasn't cool.

As was the case on that fateful day, when I stabbed my classmate in the foot with a pair of scissors.

## The battle for prestige and brawn

The victim was a member of a rival gang who wanted to prove himself to his pals. That's why he chose me — one of the most notorious boys in the school — to steal breakfast from. His booty: a brawn sandwich. If it had happened now, he would probably have stolen my phone, but back then, in the late 1950s, meat sandwiches were the measure of all things. And top of the list was brawn — known to

Americans as headcheese — and its more-vinegary version, souse, which were popular in those meagre postwar days for their high fat content and meaty flavour.

My reaction was spontaneous, but, looking back, it is hard to say whether the theft of my brawn breakfast was the main motive, or the presence of my fellow gang members judging my every move. Both were extremely important to me. Anyway, I immediately saw red, grabbed a pair of scissors from my satchel, strode over to the brawn thief's seat, bent over him, and — just as he was taking a hearty bite — stabbed him with the scissors. The kid got a hole in his foot, and I was dragged off to the police station.

Now, everyone thought I was even tougher and cooler than they did before. But, in fact, I was nothing of the sort. Whenever we engaged in any criminal activity, my heart would be in my mouth. I was able to hide it, or to carry on in spite of myself; and anyway, my actions were sometimes so impulsive that there was no time for fear to kick in and slam on the brakes — which will become even more apparent as this story continues.

I was marched to the police station, with a female police officer close behind me. Just before we entered the building, I happened to spot a banknote on the ground: 100 schillings! I immediately put my foot on the bill, bent down, and pretended to tie my shoelace. Unfortunately, the policewoman was observing my every move. Her reaction was probably just as spontaneous as my scissor attack shortly before: 'My God, what a psychopath!' She then muttered something about how they should lock me up and throw away the key. It was obvious what she was thinking — this lad is being arrested for stabbing someone in the foot, and all he can think about is making sure he gets his hands on 100 schillings that belong to somebody else!

In short: she saw me as extremely cool, too. Not in the sense of 'role model', but rather in the sense of 'scum'. However, putting my foot on that banknote was not the result of some cold-hearted calculation; rather, it was triggered by a spontaneous impulse. Like a dog clamping its jaws around a piece of meat it finds, I put my foot on the 100-schilling note and launched a cover-up manoeuvre. That was everyday behaviour in our gang, something we had become really good at; we had conditioned each other to act quickly and precisely to our own advantage, and then disguise or conceal it and adopt an innocent air so nobody noticed. And brains have a tendency to repeat actions they are particularly good at, without the involvement of the conscious mind to consider the possible consequences of those actions. The main thing is that the desired effect is achieved.

The fact that well-learned behaviour patterns are recalled and repeated over and over again leads to the conclusion, which is supported by statistical evidence, that violent teenagers turn into violent adults; if not always, then often. Why, then, did that fate not befall me? Why, in the more-than-half-a-century since the scissor-stabbing incident, have I never resorted to any kind of violence, not even an impulsive slap? There are several reasons, but the deciding factor was a complete change of surroundings, both in terms of location and in terms of my social environment. A different, better school, new friends, and — finally! — girls; different rules of behavioural effect and different role models. Learning and memory recall are dependent on context, as I'll discuss more than once later in this book (for example, when I examine addictive behaviour).

So, I did not get locked up; my father came to release me from police custody a few hours later. However, he had finally had enough of his 15-year-old son's antics. He threatened to force me to start an apprenticeship as an upholsterer, and even made me do some work experience in an actual upholsterer's workshop. That convinced me pretty quickly that staying at school was the better alternative. After all, as much as educationalists are loath to accept it, negative stimuli can be effective. My father and I made a deal: he would allow me to change schools, and I would finally start behaving. The agreement worked. In my new surroundings — a liberal school that, unlike the strict, Catholic rearing station I attended before, did not positively reinforce my thuggish posturing with attention and punishment — I was able to break out of the gang and eventually graduate.

I went on to study at the University of Vienna. There, however, my new-found middle-class conformity began to crumble, and I was drawn into the cheerful anarchy that was burgeoning there. Along with other young scientists, I protested against the completely outdated content of the courses being taught, and against the incorrigible high-ranking members of the former Nazi regime who still dominated the university. We sabotaged the lectures of established professors and held our own instead.

All this sometimes became pretty violent, and I'm not convinced the motivation was always political. Just like back in my teenage street gang, we rebels were keen to impress each other. And to impress the attractive female students, too, of course. It was the same phenomenon: our brains wanted to achieve effects such as social acknowledgement and attention from like-minded peers. And just like in my youth, I had a big mouth, but secretly I was peeing my pants. No sign of the psychopath, but rather a lot of fear, which I made great efforts to compensate for and to hide. And I had a great deal of impulsiveness, which got me into rather a lot of trouble.

I was thrown out of university and a rapid round of phone calls to other institutes of higher education meant I had no chance for the time being of getting a place anywhere in the German-speaking world. I went to England, and it was only much later, after the dust had settled over my Viennese exploits, that I was able to switch to a university in Germany itself.

### Memories, or made-up stories?

I have often been asked — indeed, I have sometimes asked myself — whether my childhood and teenage experiences were partly responsible for my later scientific interest in anxiety, psychopathy, and the brain's self-control mechanisms. Some have even theorised that the unusual circumstances of my birth contributed to my independence (if you see it positively) or unpredictability (if you see it negatively) as a scientific researcher. I was born in 1945, on a plane that had made a forced landing on the way from Czechoslovakia to Austria. I was immediately baptised four different times, because my father, desperately in search of care for his newborn son and his wife who was suffering from puerperal fever, went to various churches to seek help.

I believe such connections to be nothing more than speculative constructs that people like to recount and to hear because they lend to their lives a patina of consistency, as if everything followed a certain internal logic. This representational strategy is particularly prevalent among those who believe their success in life was somehow predetermined — those who believe they were 'always destined for greatness'.

The fact is, however, that we have only very patchy memories of what happened to us in our childhood. This has a great deal to do with the phenomenon of 'state-dependent learning', which means that memory retrieval works best when an individual is in the same or a largely similar state of consciousness as they were when the memory was formed. This explains why many exam candidates fail, because they are unable to retrieve information in the stressful examination situation of the examination hall, which they have memorised in the comfortable home setting of their student rooms. This is also why most psychotherapeutic treatments of severe anxieties and depression fail, since it is impossible in the context of a calm conversation to tackle the enduring consequences of a serious accident or abuse suffered by the patient.

And this is also the reason why we often deceive ourselves when we tell stories about our childhood or adolescence. Our lives in the past were very different to the present. Not only were we smaller, less experienced, and dependent on our parents, with our whole lives still ahead of us, we were also different, psychologically. As ten-year-olds, we are not troubled and confused by our sex hormones; as teenagers, very much so; and as 60-year-olds, not so much again. Which is not to say that our minds are largely dependent on sex hormones. But there is no doubt that they play a large part in shaping our life situation at any given time, and, according to the theory of state-dependent learning, this means we have difficulty consciously remembering what happened to us as six- or ten-year-olds because at that age we 'marched to a completely different beat', hormonally speaking, than we do today.

The bulk of our earliest reliable memories come from the time of puberty. Our memories of earlier experiences are concocted later, or they are related to us by those who were there at the time, such as our parents. It is, however, uncertain whether their memories are true or also concoctions, since parents, too, make up their own histories. And those histories, of course, can be dependent not only on the situation, but also on their own interests.

My protestant mother, for example, denied that I was christened four times, while my atheist father insisted those baptisms took place. It was not until many years later that I was able to verify this by checking church records.

## Situation, not personality

This tendency to make our own life story coherent in retrospect is also closely connected to the fact that we believe we have a personality, an immutable essential character. Some people even believe that there is a 'higher meaning' to our lives. There is no evidence for either of these beliefs.

The American psychologist Stanley Milgram managed to persuade 65 per cent of his randomly selected subjects — from a representative sample of healthy, non-violent candidates — to administer what they believed were electric shocks to a life-threatening level, to complete strangers. And those people included some that nobody would ever have thought capable of such behaviour. Their actions were controlled by the force of the situation they were in and the dispassionate repetition of instructions by the researcher, not by the force of their personality.

The research was replicated in many countries around the world, and always came up with the same result: completely healthy, apparently mentally and emotionally well-adjusted people obeyed authority right up to the point of killing. And these subjects had no personality traits that would have predicted their unconditional obedience — just as the 35 per cent of subjects who resisted authority and broke off the experiment showed no particular personality traits that might have predicted their refusal to comply.

We do not have an 'essence', nor do we possess an immutable character that guides us through our lives. It is rather the case that we function in a certain way, and are able to observe ourselves as we do so. Our brains are permanently checking whether our actions have the desired effect, whether they bring benefits (respect, success, wealth, prestige, love) — if that is the case, they are repeated. If that is not the case, they are quickly suppressed. This has helped us survive in the natural world. But there is no 'deeper meaning' to it.

Those who assume that the way they function in certain situations means that those patterns of behaviour are part of their personal essence, are wrong. Rather, external circumstances and happenstances play a much greater part in our lives than we would like to believe. It starts with economic circumstances — as a child from a poor working-class family, I would never normally have been able to go to university — and ends with the people we meet throughout our lives — I would probably never have gone into brain research if it weren't for the psychologist Hubert Rohracker in particular, who drew my attention when I was still a student to the way in which our behaviour is dependent of our brain, which in turn awakened an interest in me in the processes that go on inside our skulls.

There is little point in speculating about whether there is some kind of 'explorer personality' in me that always made me want to get to the bottom of things. There was little of the Captain Cook in me when I stabbed my schoolmate in the foot with that pair of scissors. And when that policewoman called me a psychopath, I just thought it was one more insult among many; my scientific interest in that subject came about much later. We 'wing it' from one situation to the next in life and in each of those situations, our brains seize on the behaviour options it hopes will bring the desired results, or which have produced the desired results in the past.

I never used to be interested in the slightest in psychopathy, and now I do not stab people in the foot. Which is not to say that there is more integrity among university staff than in a teenage gang. But if I were to stab one of my colleagues in the foot, it would cause me nothing but problems; there would be a scandal, and no one would look up to me or respect me for it. My brain no longer expects to gain the desired positive effects from such an act; it has now been trained in less brutish methods of conflict resolution. That is not because I have turned into a thoroughly moral character; it is because those actions that used to produce the desired effect now have the opposite result. Conversely, the behaviours that now result in helpful effects were unknown to me then, so I could not have employed them and therefore could not have achieved any effect from them.

## The brain doesn't mind

This brings me to two of the main, fundamental theses of this book:

1. The brain desires effects that it assesses as emotionally positive. It really does not matter to the brain what these consist of. There are some people who love to hurt others; and there are people who love to be hurt. There are people who need to be active all the time,

wherever they are and whatever they are doing; and there are others who prefer to do nothing and to keep a low profile. Passivity can also produce positive effects, even if they consist only of a reassurance for the passive person that he or she has done nothing wrong. Brain areas in the orbitofrontal cortex and the limbic system evaluate situations and events as they happen, measuring them against the gratifying or punitive effect that they achieved in the past or which is currently expected. These expectations control whether we engage in reward-seeking or punishment-avoidance behaviour. If this seeking or avoidance is successful, the appropriate behaviour is imprinted and stabilised. At the physiological level in the brain, this process involves neurohormones and transmitters that also play a central role in addiction.

2. The brain is open to anything as long as it achieves a desired effect, since an immutable personality is of little use in the fight for survival. Rather, it is necessary to be able to react flexibly to changing situations. Charles Darwin spoke of 'the survival of the fittest', by which he did not mean the strongest, but the most adaptive. Cockroaches — which have been around for 200 million years! — manage this thanks to their extremely resilient bodies; they can even survive nuclear explosions and removal of their brains. Evolution has given human beings a different strategy for secured survival: the extreme plasticity of our brains. They are able to adapt to ever-changing challenges, to reorient themselves, to open up to and adopt new values and topics.

We often criticise people for changing their minds or their preferences, for seeking out new enemies, friends, or sex partners, because this makes them unpredictable. Just consider the way 'turncoats' were criticised in East Germany after the fall of the Berlin Wall and German reunification. Such rebukes may be understandable from a moral point of view, since ethical precepts are usually aimed at preserving social values, but from a psychological point of view they are completely baseless. This is because the brain can theoretically adopt anything as its guiding principle for thinking and acting. That can be gaining power over other human beings, bringing about the destruction of the entire species, offering boundless love for God, or nurturing a passion for a famous musician or actor.

We have recently heard a lot about the 'selfish brain', which thinks only of itself, constantly craves sugar, and ignores the needs of our other organs. Even this theory ascribes a stable preference to the brain, which it does not, in principle, have. There are cases in which the brain brings about its own downfall simply by pursuing a certain effect. Addiction is an obvious example, but others are bravery, martyrdom, and heroism.

There will be those who think that my own theories go a step too far. And such an attitude is perfectly understandable, since the plasticity of the brain and its fundamental indifference mean, among other things, that human beings are unpredictable variables. There is nothing stable about the mind, and therefore also about ideology and character. It is possible that the person we have just married might make our blood boil with loathing in ten years' time. It is possible that the co-worker you find so unbearably arrogant now will in two years be leading you down the aisle. It's possible that our children might hate us when they grow up. It is possible that a passionate socialist might turn into a neo-Nazi, or that a money-grabbing plastic surgeon might turn into a self-sacrificing doctor. Saul became Paul on the road to Damascus, but the opposite process is just as possible. Facing such changes is often less than pleasant and, in some cases, can really pull the rug out from under us.

However, the completely open plasticity of the brain also means that mental illnesses such as anxiety and depression, cognitive problems such as ADHD, and degenerative diseases such as Parkinson's and dementia can be influenced — most importantly by the patient him- or her*self*. Even epilepsy and stroke patients can prepare themselves for a return to everyday life; even completely paralysed, locked-in patients can communicate again with their fellow human beings and achieve a state of happiness — if they train their brain in the right way (see illustration on page 26). The enormous degree of plasticity exhibited by the brain also means we don't have to confront illness or accidental injury with fatalistic acceptance; rather, we can learn to deal with them almost perfectly.

There was a time, in the 1970s, when many psychologists and brain scientists, myself included, believed that *everything* could be learned or unlearned by conditioning. Schizophrenia? No problem, if we just show the brain how to adjust its perceptive filters to avoid being overwhelmed by a tangled mass of signals (see illustration on page 27). Cancer? No problem, if we just teach the brain how to interrupt the flow of blood to the tumour. Some of us even believed infections could be successfully treated by training patients' brains to activate their immune systems. No more pills, no radiation therapy, and no psychoanalysts — just patients learning to reorient their brains with neurofeedback or other techniques. It seemed as if we were on the verge of a revolution in medicine and psychotherapy.

That euphoria, the kind which often accompanies the rise of new directions in science, has since dissipated. We now focus on that which is possible, and try not to overshoot the mark. We have not always succeeded in achieving that. For example, when we first discovered that we could help patients with anxiety disorders by regulating the activity in certain regions of their brain, we ventured so far with this technique that some of our patients developed the opposite problem to the one we were treating them for. They no longer feared anything, which unfortunately tended to lead to situations that were dangerous for themselves and for other people. These effects were not usually long-lasting, but they meant we needed to take a more cautious approach to the problem. After all, even researchers' brains should be plastic enough for them to be able to cast off any pretence of omnipotence before it is too late.

## SELF-REGULATION OF BRAIN METABOLISM

*(magnetic-resonance brain–computer interface,
or BMI — brain–machine interface)*

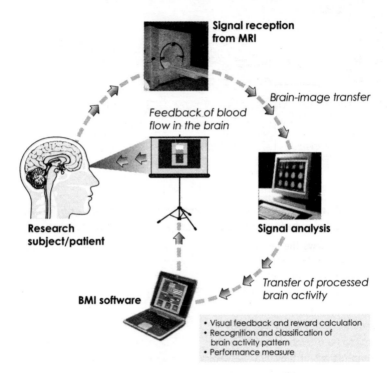

**Signal reception from MRI**

*Brain-image transfer*

*Feedback of blood flow in the brain*

**Research subject/patient**

**Signal analysis**

*Transfer of processed brain activity*

**BMI software**

- Visual feedback and reward calculation
- Recognition and classification of brain activity pattern
- Performance measure

### Configuration of a neurofeedback system for self-regulating brain metabolism

The person (left) lies in a magnetic-resonance scanner (MRI, top), which measures the blood flow in one or several areas of the brain. The person monitors her own brain activity, measured by detecting cerebral blood flow with functional magnetic-resonance imaging (fMRI), which is represented on a screen (centre) as a red or blue 'thermometer'. When the person increases the blood flow to a selected region of her brain, the thermometer shows red; when she reduces the blood flow, the thermometer's colour changes from red to blue. The computers (right and bottom) analyse the person's brain activity in real time so that she is always aware whether she has achieved the learning criteria.

## RECOGNITION OF EMOTIONS IN FACIAL EXPRESSIONS

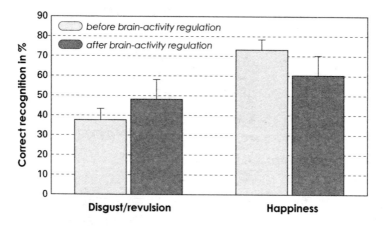

*Recognition of positive and negative facial expressions by schizophrenics before and after fMRI neurofeedback training*

This chart shows the result of what appeared to be completely harmless brain training for patients with schizophrenia, but which resulted in undesirable side effects.

Patients with chronic schizophrenia have great difficulty recognising emotionally critical, negative facial expressions in other people. My associates, the Chilean psychiatrist Sergio Ruiz and the Indian engineer Ranganatha Sitaram, trained such patients to improve the blood flow to certain regions of their brain — in particular, the anterior insula, as this area is involved in recognising negative visual expressions in others.

While inside a magnetic-resonance scanner, patients watched a coloured thermometer, which rose to red levels when blood flow was increased to the insula region of their brain (see illustration on page 26). After about ten hours of training, patients had learned to achieve this desired outcome at will. Before and after the training, Ruiz and Sitaram tested the patients' ability to recognise positive and negative facial expressions. As expected, patients were significantly better at recognising negative expressions after the neurofeedback training than before — but they paid for this improvement with a significant deterioration in their ability to recognise positive expressions. The increased activity in the emotionally negative regions of the brain presumably caused a decrease in the emotionally positive regions.

## To sum up

Personality development does not follow any inner logic or predetermined (genetic or divine) plan. Rather, it occurs in constant interaction with the environment, which opens up the way for chance to play a large part, and explains why a change in external conditions can cause alterations to ingrained behaviour patterns so dramatic that other people are apt to speak of 'a complete change of personality', despite the fact that this is, on the contrary, evidence that there is no such thing as personality at all. And although we may be convinced that we would never have administered those supposedly deadly electric shocks if we had been part of Milgram's experiment, or that we would never have become one of Hitler's zealous Nazi henchmen, we may be very much mistaken.

# 2

# Anything Is Possible?

*the plastic brain*

Are taxi drivers clever? We trust that great geniuses like Einstein were. The physicist and developer of the theory of relativity, which almost nobody really understands, is estimated to have had an IQ of more than 150. We also expect philosophy professors and multilingual UN interpreters, for example, to top the intellectual performance tables. But taxi drivers? Who ferry their fares from A to B and rail against passengers vomiting in the backseat of their vehicles late at night? Sure, they have to be able to drive a car, but who isn't capable of that? In London, they have to show knowledge of the locality to qualify for their taxi licence, but once that's in the bag, they can rely on their satnav like anyone else. And it will even let them know when the traffic is particularly heavy on a given route. That doesn't sound as if it requires a high IQ.

We know from real life that taxi drivers can be almost anybody. A former university professor from Tehran, forced into exile for political reasons. Or a brilliant musician, proficient not only on the violin and the piano, but also on the didgeridoo and the sitar, yet whose music does not attract audiences large enough for him to be able to make a living from it without topping up his income as a cabbie. It is impossible to say how many unrecognised experts, creative talents, and geniuses are on the roads behind the wheels of their taxis. But there are also the rednecks who spend their breaks snoozing with the latest tabloid scandal sheet over their face and complaining about the 'immigrants' stealing their customers. Or silent, passive-aggressive female taxi drivers who appear determined to make their passengers feel about as welcome as a dead cockroach in their morning muesli. The host of taxi drivers is extremely diverse. So, can we say anything about whether they are more or less clever than average?

No, we can't. But we can establish a different fact. And that is that you can only get a taxi licence, or a PhD, by using your brain's best feature: its extreme plasticity. Thus, newly qualified taxi drivers and potential Nobel Prize laureates have something in common: the activity inside their skulls is flexible and dynamic.

This is the conclusion reached by neuroscientists Eleanor Maguire and Katherine Woollett of University College London.[1] In a study in the year 2000, Maguire had already established that London taxi drivers have larger hippocampi (see illustrations on pages 32 and 34) than many other people. The hippocampus, an evolutionarily primitive region of the brain, owes its name to the fact that it resembles a seahorse, with a horse's head and the tail of a fish, and, just as that creature appears to transition from one animal to another, the main function of the hippocampus is to transfer information from the short-term to the long-term memory. For this reason, an enlarged hippocampus could be a sign of a capacity to retain large amounts of learned information. But the operative word here is 'could', since the hippocampus also plays a part in other functions, such as emotional behaviour (see illustration on page 32). Furthermore, the hefty seahorses inside taxi drivers' heads could have been present *before* those would-be cabbies decided to apply for a taxi licence, rather than having developed during driving. That would mean their enlarged sizes were the cause rather than the result of the drivers' career choice. Thus, the mere fact that taxi drivers were found to have oversized hippocampi does not provide a reliable indication of whether this is due to changes in their brains were brought about by their profession.

So Maguire and Woollett explored the issue further. Using magnetic-resonance imaging, they began by scanning the brains of 79 people who had begun a four-year course to qualify for a London taxi licence. Forty of them failed the course, which is hardly surprising since in just the centre of the British capital there are 25,000 streets and several thousand places of interest to be memorised. The brain scans of those failed candidates showed that their hippocampi were the same size as before they began acquiring 'The Knowledge' — as the thorough understanding of London is famously known.

However, 39 candidates passed the notoriously difficult test. And their hippocampi turned out not only to be bigger than those of a control group and the group who failed 'The Knowledge' test, but also to be bigger than before they started the training course.

## THE MAIN REGIONS OF THE BRAIN FOR EMOTIONS AND PLASTICITY

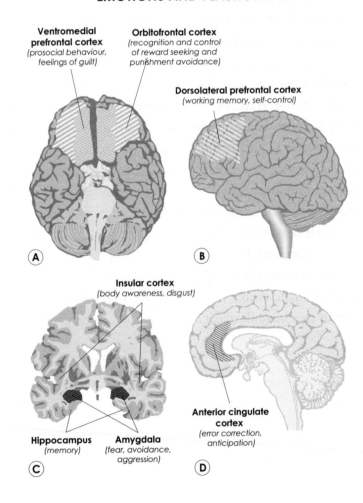

**Ventromedial prefrontal cortex**
*(prosocial behaviour, feelings of guilt)*

**Orbitofrontal cortex**
*(recognition and control of reward seeking and punishment avoidance)*

**Dorsolateral prefrontal cortex**
*(working memory, self-control)*

Ⓐ      Ⓑ

**Insular cortex**
*(body awareness, disgust)*

**Hippocampus**
*(memory)*

**Amygdala**
*(fear, avoidance, aggression)*

**Anterior cingulate cortex**
*(error correction, anticipation)*

Ⓒ      Ⓓ

The neuroscientists could not identify anything to indicate in advance which participants would complete the training procedure successfully. The drivers' brains were all more or less the same; in particular, each participant started with approximately the same amount of hippocampal grey matter. Yet after years of rigorous training, differences were apparent. The size of successful candidates' hippocampi, and only theirs, had increased significantly. They also performed better in spatial orientation and memory tests. This is a pretty clear indication that their brains had not simply changed in random ways, but in such a way as to increase their functionality.

Did those trainees who became fully-fledged taxi drivers have some biological advantage over those who failed? Could it be that their hippocampi already had a greater potential for growth? Did they have a different predisposition? Or were they perhaps just more motivated? These are all questions that Maguire and Woollett have not yet been able to answer. What all this does make clear, however, is the enormous plasticity of our brains, not only as children but also as adults. And the changes made visible by scanning techniques are just one part of this huge flexibility.

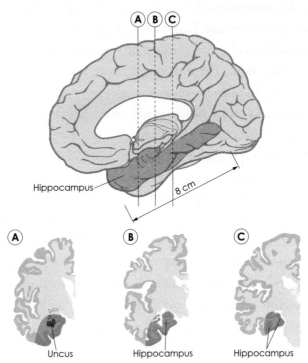

Hippocampus

8 cm

(A) Uncus    (B) Hippocampus    (C) Hippocampus

**The hippocampus plays an important part in the plasticity of the brain**

The illustration shows three cross-sections of the brain (A, B, C), at the level of the hippocampus. The hippocampus and the surrounding areas of the temporal lobe are responsible for the associative connecting of memories to the cerebral cortex. Its destruction leads to a complete loss of conscious episodic and semantic memory. This means general facts and personal experiences can no longer be remembered. The hippocampus is particularly affected by dementia.

## Stimulation and care: what really makes children's brains grow

The plasticity of our brains is apparent on several levels. Some of those levels can easily be seen using appropriate visualisation technology, and that includes the abovementioned enlargement of the hippocampus. Such processes are based on the growth of individual brain cells, and more importantly, the layer of fat which surrounds them. These days, fat doesn't have such a great image; in this fitness-obsessed age, fat is seen more as a passive mass and a dead weight than as a substance with a positive function. However, in the case of our brain cells, more fat means more electrical conductivity, and, of course, in an organ that works on the basis of the electrical transfer of stimuli, this increases its performance level.

Brain plasticity is also based on the creation of new cells and the formation of new, sturdier connections *between* neurons — an aspect that plays a crucial part in the learning process. However, these connections are an expression of not only plasticity, but also stability, since, as long as the learning processes are rewarded, those connections are not usually lost again.

Neuroscientists debate whether something once learned can ever be un-remembered by a healthy brain, since nothing is ever deleted in the brain; rather, it is overwritten, altered, and modified. A person who has learned how to ride a bicycle, swim, or play the piano but has not engaged in that pastime for the past 20 years will initially be better at the activity than someone learning it for the first time. This is because the associated neuron connections are still there, and need only be uncovered and reactivated.

However, the plasticity of the brain is also apparent within our cells, even at a genetic level. Until only a few years ago, it was believed that the genetic apparatus of our cells was fixed and always produced the same signals. Now we know that it can be altered by external influences, although only if it learned to do so early — in other words, if it gained the ability to change. Those who grow up as children in a low-stimulus environment will later lack the flexibility necessary for change — and that places a massive check on the potential for learning. Fewer connections are formed in the brain, the neurons are linked only loosely, the nourishment and growth of brain cells is reduced, and fewer cells develop in certain regions of the brain, such as the hippocampus, which is necessary for our episodic long-term memory.

The general importance of external factors on the structure of the brain has been demonstrated in experiments on rats, which were exposed either to an 'enriched environmental condition' (EC), including balls, tubes, and ladders, as well as obstacles to overcome in order to access food, or to a low-stimulus 'impoverished environment condition' (IC).[2] The results showed that the brains of the EC rats were almost unrecognisable after just a few days. They had more neurons, which were also larger overall and had larger nuclei. The number of dendrites — the branched projections of a neuron that propagate electric stimulation — had also increased significantly. Overall, the cerebral cortex of the EC rats was 10 per cent larger than that of their bored peers and was better able to regenerate following the occurrence of lesions. The cerebral cortex had not only built up more mass, but also developed a better ability to repair itself. To reiterate: all these changes took place within a matter of a few days. To achieve a similar effect in a muscle would require several weeks or months of rigorous training.

Changes to the brain can, of course, disappear and become overlaid by other learning processes, but many remain intact if they developed during a critical learning phase. Studies on human subjects have revealed how powerfully and permanently environmental stimulus causes the brain to grow. Since the year 2000, American researchers have been investigating the influence of children's environment on their development and the growth of their brain, as part of the Bucharest Early Intervention Project.[3] The researchers compared Romanian children who had lived in institutions since birth with others who first lived in such homes but were later placed with foster families, and they compared those two groups with children who had grown up within families from birth.

When the researchers scanned the children's brains, they found that those children who grew up in institutions had significantly less grey matter (made of neuronal cell bodies) and white matter (made of axons and bundles of nerve fibres) than those who grew up in family environments. Although the boys and girls who began life in institutions and were later placed with foster families had less grey matter, they had about the same amount of white matter as the children living in families from birth. This led researchers to conclude that emotional neglect and lack of stimulation causes low brain growth in children. 'Lack of stimulation' in this context means the children rarely engaged in play, received little attention, and had a monotonous daily routine. On the other hand, these deficits appear to be reversible, to a certain extent and within a certain timeframe, otherwise the children who began life in institutions and later grew up with a family would not have had the same amount of white brain matter as the original family children.

Having said this, the positive influence of the family should not be overestimated. Experiments with rats have shown that it is not so much the number of social contacts that's important for the growth of their brains, but the challenges posed by their environment. Crucially, the Romanian children growing up in institutions lacked not only a consistent attachment figure, but also stimuli; games, stimulation, challenges, and a varied daily routine. Or, to put it another way, in an environment with little stimulation, brain development is not improved by being surrounded by many peers who are equally bored.

Indeed, Japanese scientists recently discovered that overbearing parents may even hamper the growth of their children's brains. Of course, it seems doubtful that this effect is connected with boredom and a low-stimulus environment, since such parents actively seek to promote their children's development in every conceivable way and to offer them all they can. It might be assumed that the excessive fear and worry of such parents concerning their children, and their inability to grant their offspring any independence or concede them any problem-solving skills, stunts the development of those children's brains.

Kosuke Narita and his team from Gunma University scanned the brains of 50 men and women between the ages of 20 and 30 and asked them about their relationship with their parents up to the age of 16.[4] Statements like 'they tried to control everything I did' and 'they did not want to let me grow up' were interpreted as indications of overprotective parents. The team found that subjects who were 'smothered' (and 'sfathered'!) in this way had less grey matter in their prefrontal cortex (an area of the frontal lobe responsible, among other things, for self-control, emotional assessment, and memory integration; see illustration on page 32) than those who were sometimes allowed to play boisterously and unsupervised in the playground as children.

Thus, overprotective parents have a similarly negative effect on the development of their children's brains as living in a bare cage has on the development of the brains of rats. And that's not all. The Japanese researchers compared the prefrontal cortices of their overprotected subjects with those of people who had been neglected or ignored by their fathers. And they found: no difference. In other words, overprotective parents enfeeble their children's brains just as much as fathers who do not bother about their offspring at all.

Of course, such research results must be treated with caution, since people can lie or be mistaken when asked about their childhood. Furthermore, it is also possible that some subjects were simply born with a less well-developed frontal lobe and were mollycoddled by their parents as a consequence; once again, it could in principle be possible that brain development is the cause and not the effect of a particular environment.

Nevertheless, the fact remains: living in an intact family is far from being a guarantee of optimum brain development. When parents coddle and protect their children to such an extent that they stifle them with their love, leaving them no space to develop independently, they rob their child's brain of one of the necessary prerequisites for its growth.

Alongside loving care, what really counts and what really gets things moving inside the skull is exposure to situations and experiences that earn individuals social recognition and that give them an image of themselves as active agents, finding their own solutions to problems and overcoming obstacles independently by discovering connections, gaining insights, and realising that they can influence and change their environment, even if all this also entails the risk of injury or failure. Passivity and exclusive consumption, by contrast, do not generally stimulate brain growth. And situations and experiences in which people feel powerless and which they are not able to change as they would like, such as anxiety, torture, or the absence of a father, can act as negative stimuli and even prevent brain growth.

## The brain can still change in old age

'Taking up Italian at 75? Impossible!' 'Starting a new career at the age of 50? You missed that boat long ago.' There is a widespread opinion that learning is no longer possible in old age because our brains lose their ability to change and be changed as we grow older. But that is not the case. It is true that young neural structures react more quickly and dynamically than old ones, but that certainly does not mean that little or nothing is possible later in life.

In a recent study by neuroscientists at the University Medical Centre Hamburg-Eppendorf, 44 men and women between the ages of 50 and 67 were taught to juggle over a period of three months. Their brains were scanned before the three-month training period, after it, and three months after the end of the training.[5] The control group was made up of 25 people of a similar age who were not given juggling training, who were scanned on the same days as the 'jugglers'. After the training phase of the experiment, the jugglers displayed a localised, very specific increase in the grey matter in their visual-association cortex, which plays a major role in the perception of movement. The control group displayed no changes in this brain region. In addition, the researchers found increases in the hippocampus and the nucleus accumbens (part of the brain's reward system) exclusively in the training group.

This and other studies show us that the brain retains its plasticity far into old age, and that means there is practically no age limit to learning. Nonetheless, there are certain phases that are especially good for learning particular cognitive activities. For example, the phase between the ages of 18 months and three years is critical for language learning, which also places limits on therapeutic measures such as cochlear implants. These are tiny computers that are implanted in the cochlea (the spiral-shaped chamber of the inner ear) and transform sound signals picked up by an external microphone into electrical impulses. These are then transmitted via very fine electrodes to the auditory nerve. This technology allows people to (re)learn to hear, unless they lost their hearing as children and didn't receive the implant until adulthood, in which case the brain can hardly process the language signals it suddenly begins receiving. The person may develop some language comprehension, but is unlikely to be able to learn to speak normally. For this reason, it is of the utmost importance to diagnose and treat deafness as early as possible, to ensure that the auditory nerve and language centre of the brain develop optimally.

Limiting, inappropriate, and pathological patterns of perception and behaviour are also more likely to develop during certain phases of our lives. Puberty is a critical time for depression, which is often the result of separation or loss experienced during that time. By contrast, phobias, such as fear of spiders or the dark, are mostly formed between the ages of three and eight. And they develop more strongly, the more they are heeded by those surrounding the child, because attention is a kind of reward, and rewards release powerful learning impulses and strengthen the neural connections involved in information processing.

## Size isn't everything

The taxi drivers' enlarged hippocampi, the increase in grey matter among the juggling seniors and stimulated rats — these results might give the impression that learning is mainly about increasing the size of certain regions of the brain. Like a sack that bulges increasingly, the more you stuff into it. But this image does not describe the plasticity of the brain correctly at all. In the brain, it is only at the start of the learning process that certain areas become enlarged — when a person has already become an expert, this growth moves into the background, to make room for the development of other areas.

Let's take learning to play the guitar as an example. As we know, at first the learner must strive to attain a certain level of dexterity. This necessarily leads to growth in those regions of the cerebral cortex that control the motor skills of the fingers. Once those motor skills have become automatic — that is, when a learner can play without having to pass every plucking or fingering motion through the loop of conscious control — the finger-controlling regions of the cortex no longer need the space they originally required. They remain larger than those of a non-musician, but they will no longer be as large as they were when the guitarist first started learning to play.

The guitarist's movements have become automatic, which means they can be controlled by the deeper regions of the brain without any special assistance from the cerebral cortex. Instead, other parts of the brain come to the fore, such as those responsible for musical emotionality and analysis. The products of these functional shifts in the brain are what we experience as great moments in our cultural history. Adept guitarists with years of practice under their belts can 'put much more feeling' into their playing than those who are happy if they can just manage to strum a few chords correctly (see illustration on page 183).

## The brain doesn't mind — and so is open for anything

This begs the question: why should it be the areas for musical emotionality that develop when those for musical dexterity no longer require so much space? In other words, what does the brain gain from intensifying our ability to express emotions through music, once the manual skills required to produce it have become automatic? Is it hoping to gain a deeper understanding of the innermost essence of the world — in Schopenhauer's sense — which can find no better expression than in music? Or is the explanation perhaps less metaphysical and more emotional and trivial? Namely that the brain simply has more fun performing well-practised operations with feeling. Another possible explanation is that emotional expression through music earns more respect from others. This is the view usually held by social psychologists who see the brain principally as a social organ interested in provoking effects in other people.

Which of these explanations is correct is a matter of pure speculation. It is probably a combination of them all. A philosopher might focus on Schopenhauer, while more-sociable or even vain individuals might place more emphasis on the social aspects of the question. However, the question of what controls the brain's development, who or what maps out the path of its plasticity, is one that concerns more than just music. We need to ask in general terms: who or what decides the way in which a brain develops?

The answer is: nobody. Except the brain itself. Which

does not mean, however, that anything is predetermined. The brain is interested in achieving the effects it desires. It wants to accomplish, affect, set things in motion. However, *what* it wants to achieve is open. Although it is interested in what might be good for the brain, what that turns out to be depends on what the brain has learned. Initially, the brain is completely indifferent to everything going on in the outside world. This indifference only disperses once constant interaction with the environment teaches the brain what is important to it.

Depending on the nature of that environment and how the brain reacts to it, the resulting character may be egoistic or altruistic. It may be kind and affectionate or cruel and malicious. It may favour reproduction and survival of the species, but also pain and extermination. There are people who sacrifice their lives for others, and there are people who sacrifice others' lives for their own. Initially, it's all the same to the brain; in other words, each of these is just one of a myriad of possible objectives that could become a target for its interests and desires.

The extreme plasticity of the brain means that it may be aimless at first, taking a specific direction only later, in the course of a constant learning and memorisation process. So it is impossible to say that the brain necessarily serves the preservation of an individual human being, or, indeed, that of the whole human species of *Homo sapiens*; it is not automatically 'biopositive', to use a word coined by the German expressionist poet Gottfried Benn, and can take a different direction and even disengage from life altogether. Which, however, does not necessarily mean suicide — I will return to this 'cessation of the will' later.

For now, let me just note that there is nothing predetermined in the brain to tell it in which direction to develop. No morality, no biological self-preservation, no human interaction, and no God. It must find and learn about its own orientation — and the associative connections it makes, as well as pure chance, play a significant part in that process.

## Character — or just habit?

Against this background of the brain's plasticity and fundamental indifference, the question naturally arises of whether human beings can have a personality, that is, a consistent character, at all, as this seems to lead to an incongruity: if we recognise plasticity as the only stable characteristic of the brain, we exclude the possibility of an immutable character; and by ascribing a distinct character to humans or animals — which we do, even basing our entire social interaction and our day-to-day lives on this premise — we are stating that there is something that is immune to plasticity.

Interestingly enough, in studies of twins, it is those characteristics with a reputation for fickleness that often turn out to be particularly stable, while those said to be persistent frequently reveal themselves to be particularly unstable. Since the late 1970s, scientists at the University of Minnesota have been recording the differences and similarities between more than 100 pairs of fraternal and identical twins who were separated immediately after birth and subsequently knew nothing of their twin siblings. The researchers discovered that the greatest point of similarity — after intelligence — was their political orientation.[6]

A few pairs of twins included in the study were particularly remarkable. Such as: Jack Yufe and Oskar Stohr, who were born in 1933 and grew up in different step-families. Jack was raised by their Jewish father in Trinidad; Oskar was brought up by their Catholic mother in Bavaria. One twin went on to work on a kibbutz and become an officer in the Israeli navy; the other became a member of the Hitler Youth. There was an awkward reunion at the age of 21, after which neither thought it was a good idea to see the other again. Nonetheless, when the brothers met again in Minnesota at the age of 47, brought together by the twin study, they had no problem getting along with each other. On the contrary. A closer examination of their political views showed they were on the same wavelength in many areas, particularly in their espousal of conservative values and their categorical rejection of outsiders. In addition, Oskar immediately abandoned his anti-Semitic stance, and his brother never mentioned it to him again. Rather than arguing, they settled on another, common enemy: Cubans, and communists in general. From that time on, there were no differences of opinion between the two twins, who remained on good terms.

Of course, this shows that blood is somehow thicker than water; and that genes can create an undeniable degree of similarity. On the other hand, however, this example of twins also exposes a typical human fallacy:

our tendency to interpret characteristics that change little, such as shyness, pedantry, and political conviction, as permanent character traits. But on closer inspection, it often turns out to be the case that they only appear so stable because they have not been exposed to any pressure to change from the environment, as no one ever made a serious attempt to change them.

Returning to the case of Oskar and Jack in this context: one twin developed anti-Semitic views because that was 'the done thing' in the fascist environment he grew up in and because he was never confronted with strongly held opposing views. His brother, by contrast, was given a strictly religious upbringing. Hardly anyone would have thought that the two ideologically so opposed brothers could even sit down together at the same table. However, in reality, Oskar renounced his anti-Semitic views in no time at all. He was not an irredeemable Jew-hater — it was just that during his development in Bavaria he never encountered anyone who might have put an end to his anti-Semitic tendencies; indeed, he was encouraged to nurture them. After meeting his brother, however, his brain was confronted with different stimuli, with the reality of his own Jewish heritage and twin brother — and Oskar took up a new political perspective, almost like putting on a new pair of underpants in the morning.

The brain simply wants to achieve the effects it desires — in Oskar's case, ideological agreement with his twin brother — but, ultimately, it does not care how those effects come about. If he had grown up in an environment characterised by the liberal 1968 generation, it is perfectly possible that Oskar would have become a staunch communist, only to convert to capitalism if the brother he met again in Minnesota had become a business executive. If I have mastery of a musical instrument, I will want to play that instrument over and over again because that achieves the desired effect for me. Colloquially, we say it 'gives us pleasure'. And it is actually irrelevant whether I derive that pleasure from the fact of my own virtuosity or from the fact that others are listening to me — all that's important is the effect it achieves for me.

And the same is true of human character traits. If I am an extrovert, I will always tend to show that trait, because it results in a positive effect for me. By the same token, if I am an introvert, I will always tend to display *that* trait, because it results in a positive effect for me. The brain always does whatever will convey the best effect in the situation it finds itself in at any given time. If my extrovert behaviour goes down well with others, the result will be a positive feeling for me, and I will repeat that extrovert behaviour again and again — and the external and internal affirmation I receive will gradually turn me into an extrovert. Contrariwise, I will develop into an introvert if my introverted behaviour repeatedly results in positive effects. And if I annoy people with my extrovert behaviour, my outwardly oriented brain will switch down a gear and may possibly even adjust to introverted behaviour.

The supposedly immutable features of a personality are mostly just characteristics that are not influenced by stimuli for change and which can therefore easily be mistaken for being invariable. However, if and when a situation occurs in which the brain no longer achieves effects with that behaviour, it once again deploys its capacity for flexibility — and the supposed character collapses like a house of cards.

## Can anyone be a henchman?

One of the scientists who recognised as early as the 1960s that personality is not exactly a reliable constant for evaluating human behaviour was the American psychologist Stanley Milgram. His previously mentioned experiment remains one of the most notorious and controversial psychological studies to this day.

In 1961, Milgram was 27 years old and working as a young researcher at Yale when he placed a newspaper ad: 'Persons Needed for a Study of Memory … Each person who participates will be paid $4.00 (plus 50c carfare) for approximately 1 hour's time.' Participants were paid in advance and were told they would be able to keep the money, no matter how the experiment ended. Milgram then told participants that the experiment would investigate some pedagogical aspects of learning, which, like the newspaper advertisement, was a lie. In reality, the trial was a test of obedience.

The participants were seated before a bank of 30 switches arranged in such a way as to deliver ever-increasing electric shocks to another person, who was strapped into a chair in another room. The switches on the shock generator were marked from 15 volts to 450 volts. The volunteers were then told to 'teach' a list of — mostly meaningless — word pairs to the person in the next room by reading them aloud so that the person could memorise and repeat them. The experimenter instructed the volunteers to administer an electric shock to the learner every time he made a mistake, with the voltage increasing incrementally each time. Via a speaker system, volunteers were able to hear the reactions of the unseen learner in the next room. And that reaction was shocking, since the 'learner' strapped into the chair was in fact an actor trained to give the same pattern of reactions for each volunteer 'teacher' when they appeared — the shock generator was just a dummy machine! — to give him an electric shock: demanding to be released from the experiment when the shocks reached 150 volts, screaming in pain and warning he had a 'heart condition' when the voltage reached 240, and remaining dead silent after 315 volts, leaving his supposed torturer and relentless punisher with the impression that the electric shocks they are administering were now simply causing an unconscious body to jolt.

Milgram found more than a hundred volunteer subjects for his experiment. And 65 per cent of those continued the shocks all the way up to 450 volts! Even though the man in the adjoining room was screaming with pain and they must have thought that he was unconscious by the end. Furthermore, they activated the switches although no one was forcing them to do so. The experimenter, who was with them in the room, simply repeated dispassionately that the shocks would 'result in no permanent physical damage' and that 'the experiment requires that you continue'. Nothing more.

Milgram's experiment caused an earthquake in the world of psychological study, since it showed that the level of obedience to authority of someone like Adolf Eichmann is no exception, and that it is not even necessary to exert very much pressure on a considerable majority of human beings to render them so compliant that they are willing to torment and even kill innocent people. All normal people could become torturers and murderers if they found themselves in a situation where torture and murder were permitted or supposedly served a 'higher cause'. Milgram produced empirical evidence of the 'banality of evil', which Hannah Arendt had described in her book about Eichmann's trial in Israel. Henceforth, no one could any longer claim that he or she would never have collaborated with the Nazis under any circumstances.

However, the question that most concerned Milgram was whether those who turned the shock generator up to 450 volts differed in their personality structure from those who refused to comply with the experiment all the way to the bitter end. To investigate this, he invited the volunteers back, and this time subjected them to comprehensive personality tests. Milgram died early of a heart attack in 1984, but his former collaborator, Alan Elms, later reminisced that the results were absolutely disappointing: 'Catholics were more obedient than Jews ... The longer one's military service, the more obedience ... Many of these findings "washed out" when further experimental conditions were added in ...' Typical characteristics like impulsiveness, extroversion, empathy, and moral sensitivity did not have any special significance. 'In fact I didn't get significant differences on any of the twelve standard scales [for personality traits]', Elms says in summary.

This does not mean that there is no such thing as personality in general. But Milgram's studies show that the given situation also plays an important a part in behaviour, and in some cases it is the deciding factor. What's more, the fact that Catholics and those with long military careers were particularly obedient confirms that this apparently stable character trait was so strongly developed in those individuals because they had not been encouraged to overcome their obedience, but rather their obedient tendencies had been constantly reinforced. If a brain repeatedly experiences positive effects as a result of obedience — for example in a Catholic or military environment — it will also be more inclined to obedience in the laboratory setting, and send a signal to a finger to activate a switch that will administer an electric shock to a fellow human being. This is because it is real experiences and successes, as well as punishment and disappointments, rather than abstract principles and norms, which shape the brain and can continue to reshape it.

I am fully aware that this may be unwelcome or unacceptable news for those who are convinced that individuals make decisions consistently and freely. But I would like to stress that the news is not actually bad at all. The enormous plasticity of the brain also means that we can learn to live with situations that we commonly consider unbearable, and can even reshape them into sources of incomparable and unexpected happiness.

## 3

# The Coming Out of Locked-in Patients

*how the brain gives a voice to the voiceless*

There is little that stretches our powers of imagination more than the fate of patients with locked-in syndrome, whose brains are strangely detached from the rest of their body because nerve signals can no longer reach their muscles. That means they are condemned to total paralysis. Not only can they no longer walk, grasp, eat, drink, or go the toilet, but also all their channels of communication are cut. Speaking, facial expressions, and physical gestures are all impossible without the use of muscles. Patients with total locked-in syndrome are not even able to use eye movements to communicate with their surroundings. What remains for them?

The Swiss philosopher Ludwig Hohl spent many years living in a dingy basement flat, misunderstood and sometimes even unnoticed by those around him. That is far removed from the fate of a locked-in patient, but it was enough to lead the isolated poet to the insight that 'once the ability to communicate is gone, life is over'. This encapsulates the feeling most of us have when we think about an absolute inability to communicate: that it means the end of everything. Over and done! After all, humans are not trees or grass plants. We are creatures with a powerful need to communicate; we want to talk or engage in exchanges with others in some way; we want to reveal our thoughts and show others that we are who we are. But all that is impossible for those who are trapped in a completely paralysed body.

What can locked-in syndrome mean, then, other than a kind of living death, which simply requires the life-support machines to be switched off to finish the process in the hereafter?

## Can a life in a state of senselessness make sense?

The fact is that locked-in syndrome is far from a kind of death before death. To understand this, it is necessary put yourself in the place of the locked-in patient — that is, to imagine the unimaginable. This involves casting off preconceived ideas about the way this situation comes about. For example, that it comes all of a sudden. In fact, it is usually a gradual process, unnoticed at first, later becoming more obvious and increasing irreversible. Rather than spontaneous strokes leading to complete paralysis, it is more often caused by diseases that take years or even decades to develop, such as amyotrophic lateral sclerosis (ALS), multiple sclerosis (MS), and Parkinson's disease. On the one hand, this means that the grim developments slowly build up and hopes of improvement are repeatedly quashed by the body's increasing process of deterioration; on the other hand, it also means that patients have time to grow accustomed to their future. Of course, this means getting used to a catastrophic fate, but that is still different to being overcome by it without warning.

When all muscle activity ceases, patients are left lying motionless in bed, dependent on a ventilator and a feeding tube. The patient's mucous membranes dry out unless they are artificially moistened. Their eyes are closed, and if someone opens them, they see only shadows at best, as their corneas dry out over the years. Their sense of touch is similarly atrophied by years spent without grasping anything and lying in the same position. This means patients literally lose skin contact with the world. Often, they can barely feel when others touch them. On the plus side, this often means they also no longer feel any pain.

However, their ears remain functional; locked-in patients can still hear. This means they are aware when doctors and relatives — thinking the patient lying in front of them with their eyes closed is in a constant, unconscious state of sleep — thoughtlessly debate switching off their life-support machines in their presence.

Research carried out by my collaborator Boris Kotchoubey shows a third of patients thought to be in a persistent vegetative state — that is, without conscious awareness due to massive damage to the cerebral cortex — are in fact trapped in a locked-in state with an almost fully or even completely functioning and receptive cerebral cortex.[7] Kotchoubey examined more than 100 patients in Germany, and our cooperation partners in Belgium reached a similar conclusion in their study of 100 patients. This means we must assume that in my own country alone, in Germany, there are some 3,000 people lying motionless in hospital beds who are being treated as if they were unaware of what is going on around them, although they are conscious and in possession of their perceptive faculties! Many of them are forced to listen to conversations about turning off their life-support systems. Even Franz Kafka could not have conceived of a more nightmarish scenario.

Or are locked-in patients perhaps not so aware of their surroundings after all? Even if their sense of hearing still works, they are unable to react to what they hear. And if a brain can no longer achieve the effects it desires, it stands to reason that it will begin to shut down and eventually no longer consciously register the stimuli it receives from the senses. Could it not be the case that locked-in patients and those in a persistent vegetative state are actually quite similar? And that the world has not only lost them, but they have also lost the world?

## First contact

Silence. The first thing you notice when you enter a locked-in patient's room is the silence, broken only by the regular hiss of the artificial-ventilation machine. It is not like an intensive-care unit, where the atmosphere is dominated by constant coming and going, and the beeping of life-support machines and monitoring devices. Locked-in patients lie in bed apparently as stiff as a corpse, as if they had already shuffled off this mortal coil. And that's why we feel afraid as we approach one's bed.

But we approach, in spite of our fear, because the human being lying there is the person we laughed, joked, ate, and chatted with just a few years before. We take the person's hand — and then comes the surprise. That hand, which appears so pale and lifeless, is soft and warm. The room is still quiet, we still hear nothing but the hissing of the ventilation machine, but now we know: there is life here; warm, pulsating life. And perhaps this person would feel better if we broke the silence a little. Not by just talking about them as if they were lost forever, and not by reducing them to just a body that might need to be moved to another bed. But by speaking *to* them, making contact with them as an important part of our life, worthy of respect, as we used to in the past. After all, that warm, soft hand can't lie. It alone is evidence that we are not confronted with a half-dead human, but with someone who is still fully alive, and not only that, but with someone who we want to be involved in our lives? Right?

In an attempt in Tübingen to find out how locked-in patients experience the world, and to learn more about their psychological state, we began by examining their brain activity by means of a classic EEG (electroencephalogram). Sensors are placed on patients' scalps to measure the voltage fluctuations in their brains.

We then confront them with various tasks and training programs. For example, letting them hear different sequences of letters of the alphabet. As soon as patients hear the letter they want to say, there is an increase in brain activity, which causes a computer to record the relevant sound. In this way, it is entirely possible to put together complete words using the answers from the patients' brains (see illustration on page 68).

Another method consists of training patients in neurofeedback techniques in which they learn to control their own brain activity in order to pick out the desired letters. After hearing a prompt signal, they create the appropriate brainwaves by activating specific thoughts or emotions. Many people imagine making a movement and thereby create a frequency of more than 20 hertz (cycles per second) in the areas of the brain responsible for movement. Others stimulate a frequency of 8–13 hertz in the same brain regions by imagining stillness. At the same time, an audio signal increases or decreases in volume to allow them to monitor whether their brain activity is increasing or decreasing during these acts of concentration. When they succeed in increasing their brain activity at the appropriate time, they are rewarded by the computer ('Well done!'). In this way, they learn — albeit after many hours of practice — to stimulate the desired brainwave frequency alone and this allows them to select letters from a list as it is slowly read out to them, which are then repeated aloud by the computer to inform them of the result and give them a chance to correct it if it is wrong. The advantage of this method is that it does not require any muscle activity, speech, or eye movements; rather, communication takes place via a so-called brain–machine interface (BMI). This should be ideal for patients with total locked-in syndrome.

However, despite the success of this method with healthy volunteers and with locked-in patients who still have the ability to move their eyes, when we attempted it with totally locked-in patients the results were crushing. There was no indication that those patients retained any conscious awareness, let alone an ability to communicate. However, a classic EEG works using electrodes placed on the *outside* of the skull. That is, at a distance from the brain, not to mention the fact that the cranium does not let all electrical impulses through.

Therefore, we decided to use a different, much more informative method, in which the sensing electrodes are placed in numerous locations *within* the brain itself. We initially planned to perform the operation on two patients, but this confronted us with an ethical dilemma: who should make the decision about such a massive intervention when the patients themselves could — presumably — understand everything going on around them, but were unable to communicate?

Legalistically, the answer appears to be clear: the decision is incumbent on the court-appointed guardian of the patient, usually a family member, but in some cases it is the guardianship judge him- or herself. When such guardians provide written permission, the brain operation can go ahead since it counts as a medical emergency; and if the operation results in an improvement in the ability to communicate, the patient's quality of life will also improve. But does the patients share this — unverified and speculative — view? There is really no way to tell, since the very aim of the operation is to allow the currently non-communicative patient to 'talk'. Non-linguistic techniques have to be found to allow patients to express their consent or refusal.

One way is to use the mucous membrane of the mouth. We chose this method to ask the consent of a totally paralysed patient whose guardianship judge and husband had already given their consent — albeit hesitantly. We asked her to imagine drinking a glass of milk if her answer to our question was 'yes', and to visualise drinking lemon juice if the answer was 'no'. By measuring the acid content of the mucus in her mouth, we were able to ascertain whether she was signalling 'yes' or 'no'. If the pH value sank (more acidic), the answer was 'no', if it rose (more alkaline), the answer was 'yes'. We then informed the patient of the risks and possibilities associated with implanting electrodes in the brain and asked her if we could perform the operation. The mucus in her mouth signalled her consent. We repeated the question several times over the next two days and her answer remained the same. Of course, we still had our doubts, since the pH level of mucus is not the same as a spoken or written word. I spent many a sleepless night worrying about this, but eventually we decided to perform the operation.

In the case of the second patient, our decision was made easier by the fact that he was able to give his consent before he slipped into his totally locked-in state.

Neurosurgeons from the University of Tübingen then implanted electrodes in the brains of both patients. Subsequently, whenever one of these contacts showed increased activity in a desired region of the brain, we rewarded the patient, for example by encouraging them verbally. In this way, we 'seduced' them into repeatedly activating certain areas of the brain in expectation of a reward. The aim of this positive reinforcement was to turn *random* neuronal activity into *deliberate* activity controlled by the patient. We were already imagining playing recordings of spoken letters or sequences of letters to patients so that they could select the ones they intended by activating a certain area of their brain, which we would measure with our instruments. It is easy to imagine how long it takes to piece together a word or even an entire sentence using this procedure. But the concept of patience acquires a whole new meaning when you are working with locked-in patients.

And that patience is not always rewarded. We had one patient who appeared to provide purposeful yes-or-no answers to our questions. She was able to confirm the names of her children and the profession she had learned. We then went one step further and asked her questions to which we did not already know the answers. For instance, whether we should change the position she was lying in, or whether we could implant more electrodes in her brain — which she answered in the affirmative. However, over the next few weeks, all logic disappeared from her brain signals. Her yeses and noes changed as randomly as flipping a coin.

Communication was more reliable with another patient, George, who was a former soldier. However, he had had the opportunity to learn the brain language to some extent while still able to use his eyes to see. He was able to recognise a change in colour on a computer screen, or see a point moving up or down as he managed to activate certain areas of his brain. The way he managed this, what exactly he thought of to achieve it, was left up to him without any interference from us. We adhere to this principle of non-interference when using the brain–machine interface to this day. As the saying goes, thoughts are free — and why should that principle not hold for locked-in patients in particular, who consist almost exclusively of thoughts?

Another advantage for our still-sighted former soldier was that he could give conformation using eye movements, whenever we were unsure about his brain signals: 'Did you mean "yes"?' If he raised his eyes, as we had agreed with him earlier, that meant we were correct.

However, at some point, his eyes fell shut and he was left optically and communicatively in the dark. Just like the female patient mentioned earlier, his yeses and noes were now purely random. We made another attempt to improve communication with him by also implanting electrodes *within* his brain, but the results remained disappointing. Not just for us, but also for his carers and family members, who had hoped they would soon be able to communicate with him again.

## Where there's a will...?

These results, then, appeared to show that totally locked-in patients no longer have any interest in the outside world. And that would be a logical conclusion, since all movement — and thinking is ultimately nothing other than a form of movement! — aims to achieve an effect.

When a baby cries and realises that this prompts a rapid response from its parents, it will include crying permanently in its repertoire of behaviours. If, by contrast, no one responds, crying will disappear as a behaviour pattern. When we close our hand around a glass we, literally, get a grip on it; if that were no longer to be the case, our brain would no longer issue instructions to that end and our hand would remain motionless. No effect, no action. Thus, it would be perfectly absurd to hope that locked-in patients could communicate with us over the long term when their brains are no longer able to make anything happen.

An electronic 'caress' to the reward centre of the brain may be nice, but it is no substitution for a real effect. Locked-in patients may still be able to fantasise and wallow in the memories stored in their intact cerebral cortex, but their capacity for intentional thinking becomes paralysed. Their will is snuffed out — and with that, they are no longer what we would generally consider an animate individual. Such were my thoughts, at least, at that time.

However, I was not ready to give up yet. I was plagued by the unbearable thought that these people might still yearn to communicate. I was also loath to disappoint the patients' families, who had a deep desire to have contact with their loved ones and who were also unwilling to abandon all hope. After much head scratching, I hit on another method. The electrical processes of the brain had proven difficult to control. But what if we took a different approach to the BMI, and used changes in blood flow, which we could measure with near-infrared spectroscopy (NIRS), rather than electrical impulses, to allow the brain to speak?

The 'communication advantage' of blood flow lies in the fact that there are receptors and sensory organs for it in the blood vessels. Patients are able to perceive where their blood is flowing with higher or lower pressure. Even in the brain. We can feel changes in the metabolic activity and the associated blood flow. We may not be able to put it into words as we would describe something we see or hear, but we register it nonetheless. This is in stark contrast to the core business of the brain, electricity, which we cannot feel, because we lack the necessary receptors for it. Which also makes sense. Just imagine if every thought we had caused a different physical sensation, perhaps even pain! Our brain would then be completely trapped inside itself, permanently occupied with thinking and the sensations that thinking caused. It would get caught up in an endless, senseless cascade of thoughts and feelings, which would presumably overwhelm us. Evolution has therefore left us without such receptors and provided us instead with a feeling for blood flow in the brain. That's why this appeared to be more promising than an EEG as a way of enabling our patients to speak. But would it really work?

We were all eager to find out and keen to put it to the

test. So we launched another study with a totally locked-in patient. And indeed, after just a few days, she began to make contact with us via the blood flow in her brain! This, despite the fact that for months she had managed to communicate with us only very rarely via her brainwaves. We asked her yes/no questions, such as whether she would like a visit from her children, and she sent blood flowing to a particular area of her frontal lobe to signal a 'yes'. We then reworded the questions, so that the same meaning required a negative answer, and the patient had to change her original answers around. This was to make sure that there was no chance element to her responses. Our patient gave the expected responses, changing from 'yes' to 'no' and from 'no' to 'yes'. Not always, but with a certainty of more than 70 per cent over the course of many days — therefore with a much higher certainty than the 50 per cent probability that a tossed coin will come up heads (see illustration on page 68).

When the question was sufficiently important to her, the patient achieved a quota of 100 per cent. Once, she developed bedsores due to a lack of attention by her carers, and we asked her in several ways whether she was in pain. The blood flow in her brain signalled a clear 'yes'. When her bedsores had healed, we asked her, again in several different ways, whether she was still in pain. Her answers were a clear 'no'. When the stimuli were sufficiently intense — which is of course so in the case of pain — and affected her essential, vital interests, the patient was fully with us. It was possible to communicate with her more reliably than with some people who are in full possession of their communicative faculties.

This was a great success, even if successes in the scientific world are never as sudden and overwhelming and euphoric as they are on the football pitch when goals are scored. In science, they are always connected with doubt. The central question is always whether the results of an experiment can be reproduced, whether they are consistent, or whether they are just a fluke. Furthermore, the interpretation of those results could have been influenced by the researchers' anxious expectation, making them a victim of the infamous confirmation bias. This explains why scientists rarely experience feelings of triumph.

By contrast, every day and every question successfully answered by patients fills their loved ones with more relief and joy that they are 'still there' and do not appear to be in a desperate situation. And, of course, even scientific researchers are not left cold by such reactions.

Our patients could communicate! They could open up to us. They could influence their environment in order to bring about an improvement in their physical comfort. There seemed now to be no doubt: the will of a locked-in patient may weaken, but it does not disappear. And why should it? After all, these people are usually adults whose will has ingrained itself in their brain over the years and neurologically stabilised itself so strongly there that it cannot simply be snuffed out. This would be different in the case of a small child whose will is still only weakly anchored in its brain and would indeed be snuffed out if the child were to fall into a locked-in state. But in adults, it remains more or less in place.

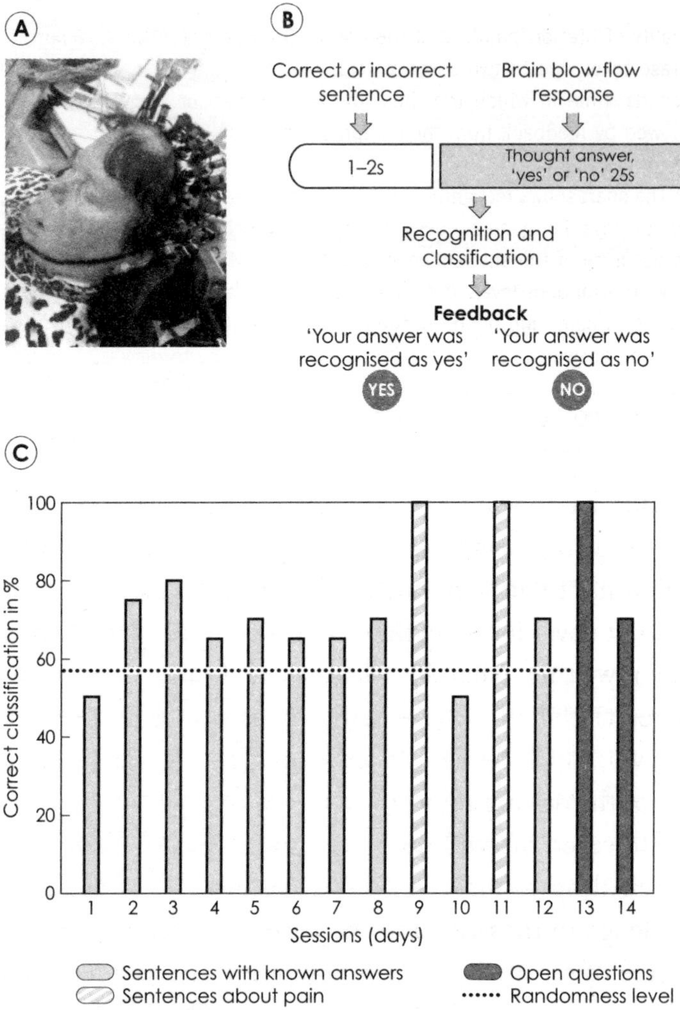
A
B
Correct or incorrect sentence
Brain blow-flow response
1–2s
Thought answer, 'yes' or 'no' 25s
Recognition and classification
Feedback
'Your answer was recognised as yes'
'Your answer was recognised as no'
YES
NO
C
Correct classification in %
100
80
60
40
20
0
1 2 3 4 5 6 7 8 9 10 11 12 13 14
Sessions (days)
Sentences with known answers
Sentences about pain
Open questions
Randomness level

***The locked-in patient Waltraud Faehnrich working on the brain–machine interface (BMI)***

(A) The patient suffers from amyotrophic lateral sclerosis (ALS). She is fed and ventilated artificially; all her muscles, including her eye muscles, are paralysed. The photo shows her with sensors attached to measure the blood flow in her brain with infrared light (near-infrared spectroscopy, NIRS).
(B) Experiment design for questions in which the key word or phrase (e.g.

'quality of life' or 'pain') is at the end of the question. This last word or phrase takes one to two seconds to say, after which the patient has 25 seconds' time in which the think the answer 'yes' or 'no'. This is then followed by feedback from the computer, 'Your answer was recognised as yes (or no)'.

(C) The chart shows recognition levels of the brain–machine interface (BMI) over 14 days. For general questions to which we already knew the answers, the accuracy of the answers generated by the BMI was about 10 per cent above the random level. Accuracy was significantly higher for questions relating to pain, and open questions. On day 9, the patient was in severe pain and answered 'yes' with 100 per cent accuracy to the question 'Are you in pain?' On day 11, she was no longer in pain and she answered 'no' with 100 per cent accuracy. The final two bars on the chart show her answers to open questions ('Are you happy to be alive?', 100 per cent 'yes').

This has nothing to do with any concept of *Homo sapiens* being metaphysically equipped with an immortal will. It has to do rather with a fundamental fact about the brain: it never loses anything, but is constantly modifying and rewriting what it has learned. Once something's there, it's there forever — and this is particularly true of our will, which develops over many decades and becomes one of the driving forces of our lives.

However, the will does need to be tracked down and loosened up to make it visible. And this is, of course, a challenge to those around the patient. It requires a great degree of sensitivity. When family members, carers, and doctors are not capable of such sensitivity and simply write the patient off, the patient's will is destined to go unrecognised.

## Attention is everything

To establish contact with totally or almost totally locked-in patients, the first thing is to find out whether they are aware. When they are dozing or sleeping, it is just as impossible to communicate with them as with healthy subjects. But when do locked-in patients sleep? Their external appearance offers no indication in this respect, since their eyes are always closed and their bodies still, the only little movement caused by rhythmical wheezing of the ventilation machine as it pumps oxygen through a hole in the patient's trachea.

Luckily, an ordinary EEG can provide more information on the waking state of such patients. And this shows us that their normal day-night rhythm has often ceased. This is not surprising, since it is no longer of any use to them. Instead, their sleeping patterns become rather chaotic. Extended periods of dormancy alternate with shorter periods of heightened awareness, which mostly do not coincide with our own — locked-in patients' periods of productive wakefulness often occur between three and five o'clock in the morning, which poses a considerable problem for researchers, who naturally prefer to work during the day and sleep at night. Not to mention the fact that even for a healthy individual who is interested in the subject matter, continuing a conversation by pressing a button, with just yes/no answers and without the aid of gestures and facial expressions or the written word, can be a rather tedious activity. Those who wish to communicate with locked-in patients must adapt to them and their particular rhythms and means of communication — and that is not easy.

This is where the patients' family members come into play, since they are ultimately the ones who can open a window on the patients' life or close it forever. The patient described earlier was lucky that her husband never left her side. He also never asked us to stop the experiments we were carrying out with his wife, seeing them rather as a chance to re-establish contact with her — a chance he intended to seize, whatever the nature of that contact might turn out to be. He learned to use the BMI so that he could communicate with his wife without outside help. Most locked-in patients are never given such a chance, since their doctors and their relatives dismiss them as hopeless cases who can only be granted a dignified end. And it is true that if nobody engages with locked-in patients, their situation does indeed become hopeless and without prospects for the future, since they have no positive experiences and no means of communicating. The premise that they should not be made to suffer unnecessarily long then becomes a self-fulfilling prophecy.

In a different case, we were helped in gaining a patient's attention by her husband. The couple had been married for many years and they knew each other better than anyone else. And so he aroused her attention by stimulating her nipples. He was also able to tell by the state of his wife's nipples, whether his wife was responsive to speech. Thus, he would gently wake her by stimulating her nipples and developed a very precise feeling for when she was awake and receptive. None of our modern, high-tech instruments were able to provide this information nearly as accurately as the patient's husband.

Many people find such erotic stimulation of a defenceless patient disconcerting or even appalling. But physical contact is a stimulus that can still reach totally paralysed patients, allowing them to feel themselves and the presence of another person. Loving family members can provide pleasant sensations for patients by caressing them, which strengthens their connection to life and helps keep them in the world of the living. It goes without saying that we need to be certain that a patient wants to be touched by the person touching her and the normal rules of intimacy must be preserved — a woman would be more likely to experience stimulation by a random carer or doctor as an act of rape, and may not be in a position to enjoy such stimulation by her husband if carers or doctors are present and watching.

With those provisos, we usually encourage the family members of locked-in patients to stroke, caress, and massage their loved ones. After all, wouldn't it be far more terrible to deny totally locked-in patients the opportunity to rediscover this means of access to the world and to their own libido?

## The return of Ariel Sharon

Even for patients in a persistent vegetative state, there can be a doorway into the inner world of their existence. It just requires someone willing to take to the trouble to open it. This is strikingly demonstrated by the example of Ariel Sharon, who died in January 2014 after eight years in a coma.

The Israeli prime minister had suffered a stroke in December 2005, resulting in several operations on his brain. An EEG showed activity in his cerebral cortex, but the stroke had caused serious damage. At that time, Sharon's doctors predicted that the 78-year-old would not regain consciousness. In April, now in a persistent vegetative state, Sharon was moved to the Sheba Medical Centre, near Tel Aviv. He was still able to breath unassisted, but that was about all. His son Gilad wrote, 'He lies in bed, looking like the lord of the manor, sleeping tranquilly ... When he's awake, he looks out with a penetrating stare.' He was fed through a tube — and showed no response to his environment.

As early as the beginning of 2006, doctors advised Sharon's family to let the former prime minister die. Based on the computed-tomography (CT) scan, the doctors were basically saying, the game was over. But Sharon's two sons insisted their father should be kept alive. He was visited every day by family members, his sons, his daughters-in-law, and his grandchildren. They spoke to him, caressed him, held his hand, or just chatted among themselves, eating home-made food so the aromas would reach his nose. They shaved him, dabbed him with his favourite cologne, let the children play noisily in his presence, sang him songs, and played the instruments they were learning to him — indeed, his hospital room was a busy place throughout the day. At some point, Gilad had the impression that his father was looking at him and even moved a finger when Gilad spoke to him. Or was it just his imagination, wishful thinking rather than reality?

In early 2013, the neuroscientists Alon Friedman (Ben-Gurion University) and Martin Monti (University of California, Los Angeles) carried out some tests on their famous coma patient. They played him recordings of his sons' voices and exposed him to various kinds of tactile stimuli — and observed the way his brain reacted, using magnetic-resonance imaging. Those scans showed: Sharon was reacting. Not always, but often. Which of course did not say anything about his state of consciousness. Nor did it say anything about his prospects for the future. But there was no doubt that Sharon's brain was displaying recognition abilities that his doctors had pronounced impossible almost seven years earlier.

Perhaps they had misinterpreted the scan images back then. Or, perhaps Sharon's brain had recovered, at least to the extent that was just possible despite the enormous damage it had suffered. With all the commotion around him in his hospital room, his brain never had the chance to degenerate completely and depart from the world forever. Instead, it was repeatedly challenged and confronted with stimuli — and since all the commotion was more or less focused on Sharon, his brain was also able to observe effects. Under such conditions, a brain does not usually switch itself off. Rather, it tries to make use of the resources it still possesses, to play the card of its enormous capacity for plasticity.

This is also borne out by our experiences with 100 patients whose condition had been diagnosed as a permanent vegetative state. We identified 30 whose brains were shown by MRI or by EEG to retain significant reactivity to the type of cognitive tests also given to Sharon. I will never forget the way one patient suddenly opened his eyes and stared at me quizzically. After five years of lying impassively in bed! This was a patient from Israel, and, after waking up, he was even able to tell of the ordeal he had suffered in the Nazi concentration camps. His doctors and his family had already requested several times that he should no longer be tube-fed and allowed to die. However, their requests had been denied, since Israel is extremely cautious about euthanasia, due to religious considerations and the terrible events of the Holocaust. This prevented him from being sent from a supposedly irreversible vegetative state into a state that would have been truly irreversible.

# 4

# No Case for Euthanasia

*the quality of life of vegetative and
locked-in patients*

Who hasn't heard of the famous arrow poison used by
some indigenous people of South America? Usually
known as curare, it is a beloved plot device of writers of
adventures stories and crime thrillers, as it turns its victims
into puppets unable to control their own movements,
paralysing their muscles, including those required for
breathing, until they eventually exhale their last. Silently,
inconspicuously. As the famous German explorer and
naturalist Alexander von Humboldt remarked during his
Latin American expedition, curare kills more stealthily
than any other drug, and, as a man of science, Humboldt
was not given to exaggeration.

We now know that curare achieves this effect by
blocking certain neuromuscular receptors, which pulls
the plug, as it were, on any kind of motor activity. Curare
victims are still able to perceive their surroundings, to

think and to reason, but they cannot move. At all. They are completely paralysed. As horrific as it sounds for the victim, for scientists this effect is highly interesting. Some researchers — including myself — have experimented with self-curarisation. Of course, it must be done in the presence of an anaesthetist to make sure breathing continues. The surprising result is that, as long as there is sufficient trust in the anaesthetists, curarisation is experienced as extreme relaxation. This is because curare paralysis prevents the muscles from communicating fear to the brain.

## The curare scandal

In the late 1960s, a news report emerged from the USA that shook the psychological world to its foundations. It appeared that learning behaviour and thinking was not, as had been believed until then, completely unconnected to the control of physiological processes such as the beating of the heart or excretion of stomach juices — meaning that both types of processes were basically the same and so could also be influenced and controlled in the same way.

The source of this revolutionary idea was the experimental psychologist Neal A. Miller, of New York's Rockefeller University.[8] He injected rats with curare to exclude any influence of movement or motor function on their learning. He then systematically rewarded certain changes in their bodily functions. For instance, if their heartrate increased in response to a certain signal, the reward centre of their brain was stimulated electrically, causing a pleasurable sensation. The alleged result was that the animals learned to increase their own heartrate without any external stimulus, purely in the hope of receiving a pleasurable reward. The same appeared to be true of other physiological changes, such as raising blood pressure, increasing blood supply to the kidneys, and modulating brainwave frequency. The rats were even taught to dilate the blood vessels in one ear more than the other, so that one took on a vaguely blue hue, while the other was white. Yes, the rats had different coloured ears — not because they had been dyed or placed with an infrared lamp shining only on one side of their head, but because they had manipulated their blood circulation at will, despite being paralysed with curare.

Not surprisingly, these results attracted a lot of attention. And since the circulatory system and the nervous system are similarly structured in both rats and human beings, the conclusion was that humans could perform the same feats. In other words, it was assumed that humans could control their bodily functions in the same way, without any physical activity being involved in the process and without the need to develop a particular behaviour. The conclusion was that these physical changes could be achieved by pure *will*, if a person wanted to receive a reward.

It seemed that the borders between medicine and psychology had been broken down. Many research scientists, including myself, began to dream. We believed it would only be a tiny step until unhealthy or pathological physical and mental processes could be corrected using concentration alone. Gastritis? No problem! All you need to do is practise relaxing your stomach and secreting less gastric acid. Anyone could treat their own depression by simply learning consciously to alter the brainwave frequency causing their depressive state. We even dreamed of starving cancer cells to death simply by teaching patients to focus on cutting off the blood supply to their tumours. It seemed as if almost every kind of illness would soon be curable just through learning, without the need for pharmaceuticals. As young researchers, we went on pilgrimages to New York to breathe the intoxicating air of omnipotence itself, and it was suggested that Miller should receive a Nobel Prize.

However, we were soon to be brought back down to earth. Not one single researcher was able to replicate the results of Miller's curare experiments. When one of Miller's assistants eventually committed suicide, suspicions began to grow that Miller had manipulated the results, and those suspicions remain to this day. Could it all have been nothing but a great big con? Whatever the truth, Miller's name was removed from the list of potential Nobel Prize candidates, and he lost his job at Rockefeller University, and our dreams burst like bubbles.

Nonetheless, for me, the episode was important as an impulse for future theories, and I am still proud to have been a student of Miller's. Ironically, that important intellectual impulse did not bear fruits until I was forced to take my ideas in a different direction. We wanted to establish contact with our locked-in patients, who lie in their beds completely paralysed like curare victims, and Miller's failure meant we had to develop a different strategy, an alternative to deliberate reward-based learning. After all, why would completely paralysed people become active and communicate with us, when they were no longer able to achieve any effect? If rats didn't do so, why should human beings, with the ability to reflect rationally on their actions, be any different? I came upon the idea that the answer could be found in the work of a classic behavioural researcher: the legendary Ivan Pavlov.

## From reflex to volition

This Russian psychologist discovered that just the ringing of a bell is enough to make dogs salivate if they have been conditioned to associate the sound with the presentation of food, thanks to the repeated occurrence of the auditory stimulus at the same time as the dogs were fed. After a certain number of repetitions, the formerly neutral stimulus alone was enough to trigger salivation in the dogs. Pavlov called this a conditioned reflex. It is important to note that this was a *reflex* and not an act of volition.

We decided to make use of this mechanism with our locked-in patients, by asking them questions that did not require deliberation or decision-making, but which could be answered spontaneously. Examples of such questions included: 'Is Paris the capital of France?' 'Is Paris the capital of Germany?' 'Is up the opposite of down?' The answers to such questions can be given reflexively as 'yes' or a 'no', without the need for further consideration. During this process, we used near-infrared spectroscopy to observe what was going on in our patients' brains. A computer program then used these observations to create a pattern to be used in turn as a basis for comparison for later experiments. In other words, we taught the computer program to recognise the brain activity of patients when they answered 'yes' or 'no' (see illustration on page 68).

The next step was to confront patients with question that they could not answer reflexively. For example: 'Would you like to see your children?' And then the crucial question: 'Do you want to die?' The computer recognised whether patients answered 'yes' or 'no', after which we turned the question around, as a control ('Do you want to live?' 'Do you want your children to stop visiting you?'). Communication was successful, and we were able to recognise answers reliably.

Although it did not work all the time, the overall implication was clear: locked-in syndrome does not necessarily entail a total departure from life. Patients could indeed communicate if they were released from their prison in a systematic way. However, this did not address one important issue: that of the will to live, of a satisfying and fulfilled existence. It may well be a welcome change for locked-in patients confined to their curare cage to re-establish contact with the world. But is that enough to make a person happy?

## Can it be called a life?

Anyone who tries to imagine what locked-in syndrome must be like for a patient cannot but shudder at the thought of suffering such a gruesome fate. Let's say you have suffered a stroke and have just been taken to hospital.

There you are, lying in a closed room, staring at the ceiling. In the coming weeks and months, that ceiling will become your constant companion. Friends and family come to visit you, but many of them are struggling to cope with the situation, such that you are relieved when their visits are over. You are left alone with your thoughts. You think of the sex that you will never have again, of your mother's delicious beef roulade that you will never eat again, of that after-work beer you'll never drink again. All that remains of the real world for you is the hospital room around you: the soulless beeping of machines rather than the joyful wagging of your dog's tail; the smell of disinfectant rather than the scent of your favourite perfume.

And no doctor tries to ascertain whether you are still aware of anything, as that isn't normally part of a hospital's standard bank of tests. The hospital staff in general are not much help to you. They can be divided into four groups: the businesslike ones, who go about their work with indifference; the cynical ignoramuses, who think you are just a malingerer; the rough, inconsiderate ones, who believe you can't feel anything anymore anyway; and finally, those who are genuinely interested in your fate and whose efforts are, on the one hand, like balm for your soul, but, on the other hand, increase your desire for normal conversation to an unbearable level. In any case, you no longer feel that you are being treated like a fully-fledged member of the human community.

Perhaps the idea passes through your mind — many things pass through your mind at this time, since your mind is the only part of you that remains functional

— that all this is a blessing in disguise. After all, you could have suffered the same fate as the Belgian man Rom Houben. It was 23 years until he was diagnosed with locked-in syndrome! For all that time, he was presumed to be comatose and in a persistent vegetative state, unaware of anything or anyone around him. Very few patients who are assumed to be comatose or in a persistent vegetative state, or, in more modern medical parlance, with 'unresponsive wakefulness syndrome', are examined as closely as they would be by Steven Laureys in Liège and me and my team in Tübingen. That alone is a scandalous state of affairs. But fortunately, you are spared that fate, because you are attached to an EEG monitor just days after your stroke and are confronted with senseless statements like 'fingernails grow on trees' or 'elephants have wings', which provoke your logic-loving cerebrum into firing off electronic flashes of indignation.

But you are unable to draw much comfort from the story of Rom Houben. After all, even when your condition has been diagnosed correctly, your life remains bleak, with little stimulation. Not to mention the fact that you have already been told that you will never return to your old state of health. And that you will require constant care for the rest of your life — eternally dependent on the help of others; a burden on all your loved-ones. Happiness is a thing of the past. All that remains for you is agony and despair. Or not?

## The meaning of life beyond iPhones and marathons

This or something similar is the way most healthy people imagine the experience of total paralysis. It is inconceivable for them that anyone could experience contentment or happiness in such a situation. But is it really so impossible? And how can we ever find out?

Sociologists and psychologists have developed certain tests to sound out various 'happiness factors', as a way of measuring individuals' level of satisfaction with their day-to-day lives. They include questions such as 'Do you have a lot of friends?', 'Do you enjoy eating out with friends?', 'Are there things in life which make you happy?', 'Do you like getting up in the morning?', and 'Do you enjoy sex?'. The answers are analysed using a points system to provide an evaluation of the test person's general quality of life. It works in a similar way to an intelligence test and — provided the test person does not lie — has great informative value.

For locked-in patients, however, many such questions lose their meaning, because those patients are no longer able to move. They may be able to indicate whether they have a lot of friends or not, but they are not even able to eat, let alone go out to eat with their friends. Sex is usually also a thing of the past, and even getting up in the morning is an impossibility since they are unable to even sit up without assistance. After a time, their brain stops wanting to move, or at least stops wanting it so much, and questions about movement then no longer have the same significance to locked-in patients as they would for a healthy person.

For this reason, we came up with a new catalogue of questions, which do not focus on movement and which can be answered with equal relevance by both paralysed and able-bodied respondents. The questionnaire includes statements like 'I have good friends I can rely on' and 'On the whole, I am happy with my life', which can be answered simply in the negative or the affirmative. The answers given showed that both groups have about the same quality of life. In the case of some paralysed patients, the result was significantly higher, although for others it was also significantly lower than for the healthy respondents. This depends on the current stage of their condition.

Patients with amyotrophic lateral sclerosis (ALS) often feel very unhappy in the early stages of their illness, and many are plagued by thoughts of suicide in this period. One reason for this is that the end of their normal life as a mobile human being is still recent and raw; another reason is their fear of what is to come. And this fear has been found to be especially strong in doctors who are diagnosed with ALS. This is presumably because, while doctors may be aware of the way this muscular paralysis leads to the physical decline of the whole body, they never previously considered how to deal with this challenge mentally.

The mind's potential for coping in this way is huge, which is why many ALS patients achieve an impressive level of life quality as their disease progresses. They do experience a difficult crisis period when their lungs begin to stop functioning — they repeatedly experience feelings of suffocation, and eventually they need to be ventilated mechanically through a tube. This can lead to a resurgence of thoughts of assisted suicide. But once they have passed that hurdle, their quality of life can once again stabilise at a high level. This happens when such patients learn to adapt their priorities to their physical condition and let go of the priorities they had before they became paralysed.

As their condition progresses, the focus of their attention falls increasingly on those they share their lives with and on social interaction, while the pleasures of the able-bodied, such as going out or engaging in sports, begin to pale. The people they are close to, usually family members, become increasingly important, and this is why it is crucial that those people do not break off contact with the patient. People who are totally paralysed may no longer be able to run marathons or even swipe their fingers over an iPhone screen, but life still provides them with enough moments of happiness. They will 'say' such things as 'I feel happy whenever my husband comes in' or 'My biggest joy is when my son comes to visit at the weekend'.

It is perfectly possible for a severely paralysed person to lead a worthwhile life once they have mentally accepted their physical condition — and it is very likely that they will do so. Our tests show that most such patients do very well in coming to terms with their situation and achieve a high quality of life on the basis of what is still possible for them. And these people include many who believed categorically when they were still healthy that this would never be possible for them to achieve.

## Signals of happiness from the depths of the brain

Nevertheless, we came in for some hefty criticism for our findings on the quality of life of such patients. It was suggested that our patients had only reported having a high quality of life out of a desire not to disappoint us, the research team. The proposed explanation for this was that we and our machines were, so to speak, all they had left in the world and they did not want to jeopardise that. The reality, according to those who spearheaded this psychoanalysis-based criticism, the patients were weary of life and were just concealing the fact — albeit mostly unconsciously, since most fears are fed by the unconscious mind — so as not to lose that final point of contact with their surroundings. In other words: our findings on our patients' quality of life were worth about as much as a confession extracted under torture from an inmate at Guantanamo. Neither statement originates in free will, but under external pressure.

We did not waste time on considering why a person with an incurable condition should be more likely to lie about their quality of life than a healthy person who had much more to lose by comparison. However, since we doubted that deductions made without ever having communicated seriously with severely paralysed patients could lead to knowledge of their 'true' motives, we set about measuring the life quality of our patients in a scientific way.

To do so, we used magnetic-resonance imaging, because that technology allowed us to see changes in blood supply to the deeper regions of the brain. Patients were placed in an MRI scanner and confronted with images (for those who still had use of their eyes) or sequences of sounds (for those who could only hear) intended to provoke positive or negative emotions in them. Members of a control group of healthy volunteers were subjected to the same procedure. Prior to the start of the experiment, members of the locked-in patients' families were asked whether it was acceptable to subject them to the emotionally charged sequences of images or sounds.

The images and sounds were taken from the International Affective Picture System (IAPS), developed at the University of Florida by our friends Peter Lang and Margaret Bradley. This material is used by psychologists all over the world as a standardised tool for studying and recording the way emotions are triggered, and it includes images that would affect anyone in the world more or less intensely, irrespective of their cultural background. The set contains images of, for example, a pile of human skulls, maimed children, and dismembered human corpses to provoke reactions of fear, horror, and disgust in anyone who sees them. There's images of naked women (for male subjects) and laughing infants (for female subjects) to provoke feelings of lust or joy. The IAPS consists of hundreds of images, which have been tested on countless numbers of people and have been classified on a scale of one to ten for their emotional effect. For subjects without visual perception, there is a set of corresponding sounds, including the sound of the sea and of children's laughter, as well as human cries of pain and the screech of a circular saw.[9]

When we exposed our patients and the group of healthy volunteers of similar ages to the pictures or sounds, we found some remarkable differences in their MRI scans. The locked-in patients reacted more strongly to positive stimuli and less strongly to negative stimuli, and this tendency was particularly noticeable in patients who had been ill for some time and were reliant on artificial ventilation. There was no indication at all that their negative physical condition left them able to react only weakly to their environment. Contrariwise, there was also no sign that their extremely low-stimulus life caused them to be shocked or to over-react, which might have been expected. Rather, that which makes us happy made these highly restricted people even happier, and that which makes us unhappy affected them much less than us. Ultimately, the conclusion must be that their quality of life is higher than ours.

Subjects' reactions to the IAPS stimuli show up in MRI scans in different areas of their brains, depending on which signalling pathways are activated. For instance, images of happy, friendly faces affect mainly the supramarginal gyrus, a part of the cerebral cortex found in both hemispheres where the parietal, temporal, and occipital lobes meet. When this area is active, it sends blocking signals to the amygdalae and other parts of the neuronal system involved in reacting to danger. Those blocking signals cause us to relax and make a friendly face. This is a very useful mechanism for a communicative species such as human beings, as it helps us reach an emotional consensus. And this appears to be all the more so for human beings who are extremely limited in their ability to communicate, because this lends more significance to each individual communicative signal.

Our patients' supramarginal gyri reacted significantly more strongly when they were shown images of good-humoured faces than those of the healthy control group, which means our patients' brains switched more strongly to a state of positive relaxation. They were more ready to put on a friendly face — even though the paralysis of their facial muscles meant they couldn't show it.

All this raises the question not only of why people who can barely do anything at all should continue to react to their environment, but also why they should have a more-than-averagely positive attitude to life. Shouldn't we expect their brains to just give up?

## What goes around ...

To answer this question, we looked more closely at the actual lives of our patients. We found that it was particularly those who were in an environment where they felt well cared for — that is, those with a good relationship with their carers and their families and friends — who felt the most joy for life. To put it another way: locked-in patients who are surrounded by friendly, caring, and empathetic people feel good. Their brains are ultimately mirrors of their environment, just like everybody else's.

Someone who spends the whole day with malicious people will eventually have a brain that is tuned to maliciousness. Alternately, the brain of someone who spends the whole day with communicative, friendly people will be calibrated for friendly interaction. This is true of every human being, and the effect is all the greater for those who are dependent on the help of others and are able to communicate only very little due to a severe disability.

This stabilisation of, or even increase in, a patient's quality of life can be observed not only in stroke survivors, but also in those with other conditions that cause the brain to lose control over movement and other important functions. These include ALS, Parkinson's disease, epilepsy, paraplegia caused by injury, and advanced dementia. Even patients with lingering physical diseases that lead to infirmity, such as AIDS and rheumatism, often display an impressive lust for life — as long as their pain can be controlled — and show no sign of resignation. Except for some forms of cancer that cause brain tumours that disrupt the hormonal balance of the brain to such a degree that the sufferer's mental state cannot be stabilised, there are virtually no very severe diseases that prevent the sufferer from achieving a high quality of life.

It is not uncommon for people to gain strength precisely through becoming ill to make radical changes to their lives, such as ending a destructive relationship, quitting a frustrating job, or taking that trip of a lifetime they have always dreamt of. Once they have mentally processed their diagnosis and become accustomed to their altered life conditions, many people feel energised and optimistic. We are not talking about a handful of people here, but millions of human beings whose lives often may not seem worth living from the point of view of a healthy person, but who themselves feel their lives are very worthwhile. This insight should always be at the centre of any discussion about living wills, assisted suicide, and euthanasia.

## Living wills: enticement to suicide

For many years, living wills (or as the medical profession often terms them: 'advance healthcare directives') led a shadowy existence. It was seen as taboo to think about that 'worst-case scenario', when decisions would need to be made about how and if we should be treated if serious illness leaves us no longer able to communicate. But those times are over in an increasing number of countries around the world.

Since the corresponding legislation was passed in my country, every citizen has the right to determine in writing 'whether he consents to or prohibits specific tests of his state of health, treatment or medical interventions not yet directly immanent at the time of determination (living will)' (Section 1901a of the German Civil Code). The Humanist Association of Germany estimates 12 million people throughout the country have now signed living wills as a result of that legislation. Some of those may not be legally watertight, but the trend is clear: citizens want to decide for themselves in advance whether everything medically possible should be done to keep them alive if serious illness leaves them unable to express their wishes, or if treatment should be stopped and they should be allowed to die. This sounds like progress and self-determination — but in fact it is essentially the opposite.

How can there be talk of a freely taken decision if a living will signed by a healthy 40-year-old man is applied 30 years later only because he has been robbed of his ability to revoke it — due to advanced Alzheimer's disease or ALS, for example — although he is actually quite satisfied with his life? Indeed, the time between signing the living will and its application need not be so long: under German law, even seriously ill people can usually sign a living will.

Thus, in the knowledge that their illness will eventually lead to locked-in syndrome, many ALS patients sign a living will stating that they do not want to be artificially ventilated. When their breathing becomes progressively difficult, and their condition therefore becomes increasingly distressing, they insist on ending their suffering and urge the medical staff treating them to help them do so.

In Tübingen, we usually refuse such requests. We do not do this out of high-handedness or lack of sensitivity, but because — unlike the patients in question — we know how great the chances are that they can still attain a high quality of life. This is because their brain will adapt to receiving only very few external stimuli, and, by the same token, those stimuli will be experienced as particularly intense and positive, as long as the patient is well cared for. We explain this to our patients in great detail and ask them to place their trust in us and to have a little patience so that they can live to experience this effect themselves. If they agree, we sometimes even tear up the actual living-will document.

Our patients can, of course, rely on us to support their rights and follow their wishes if their love of life should ever be irrevocably lost — even without a formal declaration. Having said that, such a case has never occurred.

## Football and skiing to the very end

How wrong-headed a living will can be is illustrated by the case of one of the first patients we managed to teach to communicate with us using his brain, despite being paralysed by ALS. His name was Hans-Peter Salzmann, and, during his career as a judge in Stuttgart, he had gained a reputation as a reliable person who stuck by court decisions once they were made (see illustration on page 98). After receiving the devastating diagnosis of ALS, he put pen to paper and with characteristic meticulousness drew up a living will, which he put away in his desk drawer. The will included his wish not to be artificially ventilated.

A few months later — he was still being cared for at home — he was suddenly unable to breathe. His carer called the emergency doctor, who immediately put Hans-Peter on artificial ventilation. This was against the patient's wishes, but the doctor knew nothing of the will expressing those wishes, which remained undiscovered in that desk drawer, and Hans-Peter was no longer able to communicate.

After he was admitted to a clinic, he heard about us and our brain–machine interface, with its possibility of enabling almost completely paralysed people to communicate with their environment. Hans-Peter decided to seize this opportunity. However, when communication was successful and he was even able to dictate entire sentences via the BMI, he did not mention a single syllable about his living will.

For eight years, he not only remained alive, but also enjoyed taking an active part in that life. His favourite pastimes in that period included watching football and downhill skiing on the television. He had been an active football player and skier earlier in life, but we had assumed that a totally paralysed person would gradually lose interest in such things, since their irreversible state of inactivity would destroy the associated motion patterns in the brain. However, our judge from Stuttgart remained faithful to his favourite sports.

## A smile on a frozen mask

Hans-Peter left us in no doubt that he considered his life worth living. We watched sports shows together and, in the notes he would dictate via the BMI, he would even invite us to parties, where we would use a funnel to pour a glass of wine into his enteral feeding tube. On those occasions, we all had the undeniable impression that his face took on a happy and relaxed expression. Which was surprising, as the paralysis of his face muscles should really have made that impossible. The faces of ALS patients are usually frozen and mask-like, and so, after thinking about it rationally, we concluded that we must have imagined the change in Hans-Peter's facial expression.

However, we later realised that our first impression

was probably correct, after all. Human facial expression also involves muscles that are not subject to motor control in the brain — these also include those muscles involved in causing the hairs on our skin to stand on end when we get goose bumps — and which can therefore still function in totally paralysed patients. Moreover, those muscle movements can sometimes become particularly visible in patients with ALS, precisely because their other, consciously-controlled muscles atrophy and no longer dominate any facial expression.

As well as this activation beyond any motor control, another feature of the involuntary muscles of facial expression is that they are involved not only in expressing emotions, but also in the opposite effect of initiating, or at least reinforcing, emotions in the brain. The principle: smiling outwardly creates stimuli that also make us smile inwardly! It is not for nothing that some behavioural therapists recommend that their depressed patients somehow try to put a smile on their faces, as this will also affect their mental state.

Therefore, we must assume that Hans-Peter not only looked particularly happy after his glass of wine, but that he actually *was* happy. Later, we saw the same effect with a Peruvian ALS patient, although it was whisky rather than wine that his wife funnelled into his feeding tube.

Apart from this effect, Hans-Peter of course communicated less via his facial expression and more via the brain–machine interface, by activating specific regions of his brain to create impulses that a computer program translated into syllables (see illustration on page 98). As he was still able to see, he was also still able to recognise letters, syllables, and words on a computer screen. If he wanted to select a certain letter, he would have to create a slow brain potential — a technique he had learned and practised. If he was successful, a series of letters was split in the middle, and if the desired letter was in the remaining group, he would again have to create a slow brain potential, and so on. Hans-Peter mostly imagined sports-related actions to control his brain potentials. Later, he even learned to turn on the brain–machine interface independently, by reacting to its background ticking sound with a specific brain potential.

In this way, it was possible to hold conversations with Hans-Peter, and, after a certain amount of practice, he was even able to write letters. Contact with my collaborator, Andrea Kübler, who visited him every day, was of vital importance for this success. Furthermore, he had the same carers looking after him for many years, who were therefore able immediately to recognise and deal with any difficult situations, such as mucus blockage in his windpipe.

Naturally, Hans-Peter's story became a sensation. He was the first person to learn to communicate directly using the patterns of activity in his brain. His achievements were reported and honoured in the scientific journal *Nature*.

DEAR-MR-BIRBAUMER-
I-HOPE-YOU-WILL-COME-AND-VISIT-ME-SOON-
AFTER-YOU-RECEIVE-THIS-LETTER.-I-THANK-YOU-
AND-YOUR-TEAM-AND-ESPECIALLY-MS-KÜBLER-
MOST-SINCERELY-AS-YOU-HAVE-ALL-TURNED-
ME-INTO-AN-ABECEDARIAN-WHO-OFTEN-GETS-
THE-LETTERS-RIGHT.-MS-KÜBLER-IS-AN-EXCELLENT-
MOTIVATOR.-THIS-LETTER-WOULD-NEVER-HAVE-
EXISTED-IF-IT-WEREN'T-FOR-HER.-IT'S-A-CAUSE-
FOR-CELEBRATION-SO-I-WOULD-LIKE-TO-INVITE-
YOU-AND-YOUR-TEAM-TO-JOIN-ME-TO-DO-SO.-I-
HOPE-WE-CAN-ARRANGE-A-DATE-SOON.
ALL-THE-BEST.
YOURS-
(patient's full name)

**Locked-in patient Hans-Peter Salzmann training on the BMI**

Above is the world's first letter written using the electrical activity of the brain. In it, the patient invites the author and his collaborators to a party to celebrate his achievements.

## Living wills and the power of fear

At that time, we were still caring for our patients in their homes. Today, most are cared for in the well-equipped 'Haus CERES' centre, near Tübingen. Over time, a friendly and trusting relationship usually develops between us and our patients. This was true of Hans-Peter. One day, he told me about his living will, which was still in his old desk drawer, and I asked him if he was happy that no one had found it back when his condition deteriorated. His reply: 'Even judges sometimes get it wrong.' He clearly hadn't lost the ability to laugh at himself.

The end came while Andrea Kübler and I were away and therefore unable to care for him personally. In fact, his condition seemed stable enough not to be a cause for great concern. But our patient contracted pneumonia and was taken to a general hospital, where he was treated as a half-dead coma patient in whom they did not see any point in investing anything, either in terms of medical treatments, or in terms of care. Hans-Peter developed bed sores and died within two weeks, without having exchanged a single word more with anyone.

Hans-Peter's death is all too typical and certainly no rarity, although there is no statistical information on this issue. Young doctors in particular are often unable to recognise that a totally paralysed patient has any quality of life and — like many other people — are more inclined to think of 'putting an end to their suffering'. The consequence of this attitude is that they avoid contact and interaction with such patients. Their lack of training in dealing with such patients means they are not able to examine them objectively. Dangerous infections and breathing difficulties are ignored, and this is true not only of most doctors, but most nursing staff, too. What both professions have in common is that they are trained to cure patients, even though modern medicine is increasingly about treating people with chronic, incurable conditions, whom medical staff should not be aiming to cure, but to provide with as high a quality of life as possible despite their illness.

Hans-Peter's death was as unnecessary as it was tragic. It would not happen under our care now, as we now make sure that we are informed of any changes to our patients' state of health. But after Hans-Peter died, our only comfort was that he had at least had eight years of worthwhile life after he required artificial ventilation, which he would not have had if the wishes expressed in his living will had been followed.

This experience is one of the main reasons we now try to reach an agreement with our ALS patients that there should be no hasty withdrawal of life support. In fact, we try to agree on a waiting period of up to a year following the start of artificial ventilation. Patients who still wish to die after that period can rely on our help. In the case of ALS patients, this means they are given benzodiazepine and a dose of morphine to make them fall asleep peacefully. This is legal in many countries, including Germany, but is practised in only a few clinics, doctor's practices, and care homes. Often, the ventilator machine is simply switched off or artificial feeding is stopped, which ultimately means that they die a wretched death from suffocation or starvation. This can only be described as inhumane and brutal, irrespective of the condition the patient is in. Nonetheless, many people give their express permission for this in their living will, as Hans-Peter originally did. This presumably means they did not properly understand the document they signed.

This once more makes clear that the problem with living wills is that they are usually signed without the requisite knowledge. At the moment of signing, a signatory to a living will cannot know how he or she will feel at the time the will comes into force. Just as he or she cannot know how it feels to starve to death or suffocate when totally paralysed. People allow themselves to be taken in by the idea that rejecting all life-support measures in advance is a free and independent decision on their part.

This is a fallacy of breathtaking proportions, a confusion of limitless freedom with limitless stupidity. Freedom to decide can only come if we know what the options are — in this context, what the alternative to switching off the machines is. But that is not the case by far!

Most of us have probably heard about patients left paralysed by a stroke and perhaps even seen with our own eyes how they require constant care and have no control over their own body. But who knows what's really going on inside their heads? And has anyone ever seriously considered the possibility that such people might actually be happy and feel that their lives are fulfilling? There is a huge gap in people's knowledge here, which each individual should seek to redress for him- or herself. However, most people prefer to go with their spontaneous fears and sign a living will, which may turn out to be more of a curse than a blessing in the future.

Very few people, both within families and within the medical profession, are aware that it is possible to communicate with totally paralysed patients via the BMI. Despite this lack of knowledge, health-insurance companies have never refused to cover the cost of this procedure in any cases that I know of. Those costs amount to about €30,000 at the time of writing, but they are falling every year. There is much industrial research, particularly in Germany, aimed at producing affordable, wireless BMI devices.

Unfortunately, my institution is the only one to have staff trained in BMI procedures, a state of affairs that must be addressed urgently. This brings me to the crux of the matter. Any progress in the field of BMI ultimately depends on whether doctors and family members *want* locked-in patients to communicate with them and to return to the land of the living. As long as the prevailing attitude is that it is better to 'switch off' such patients, to 'put them out of their misery', progress in developing BMI technology and techniques, as well as efforts to lower the costs, will continue to stagnate and may even come to a complete halt. And then it will be no use looking abroad for help.

## More relaxed laws cause more fear

In countries such as Belgium, the Netherlands, and Switzerland, and in some US states, pseudo-liberal legislation has meant that euthanasia has become practically routine. It is sometimes even requested in cases of depression. Some would like to see similar regulations in Germany. But I can only warn against this, because I believe we are justified in our suspicion that relaxing euthanasia laws simply reinforces patients' negative expectations when it comes to their future quality of life. The message they receive is that if everyone is calling for legalised euthanasia, the alternative — for example, life with advanced Alzheimer's or locked-in syndrome — can only be terrible.

In Germany, by contrast, where reaction against the lasting legacy of the policies of the Nazis means legislative attitudes to euthanasia are relatively restricted, such patients do not have such a negative view of their future at all. When we spoke to ALS patients about the issue of assisted dying, only three out of 100 respondents said they had seriously considered it and had already broached the subject with their doctor. Clearly, their fear of a miserably low quality of life in the future had been so great as to make them want to put an early end to their life. Yet even those three had decided against it in the end.

The decisions people make are often driven more by fear than by rational insight. As Nietzsche put it, 'fear is the mother of morality'. In Nietzsche's time, there was no such thing as a living will, and, even just 20 years ago, barely anyone was keen to sign one, because the fear of death was greater than the fear of suffering or of losing control of life and limb. The main thing was to stay alive. This is not the case today — even in Germany, where the movement for the relaxation of euthanasia laws is making itself ever more keenly felt.

In January 2013, the German health insurer DVK published the results of a representative survey in which almost 60 per cent of 45-year-olds said they would rather be dead than suffer dementia. It seems reasonable to assume that a similar proportion would give the same answer in the case of locked-in syndrome, which is also tied up with the fear of losing independence and self-determination. Living-will and euthanasia legislation has nothing to do with moral progress or increasing personal responsibility, and certainly nothing to do with an increase in knowledge; it is nothing more than an irrational displacement of people's fears. And this attitude feeds off the fact that we massively underestimate the brain's resources and its plasticity — and therefore its abilities to heal itself.

## 5

# Hope for People Suffering from Epilepsy or Stroke

*our brain's enormous powers of self-healing*

Peer Augustinski was a popular television personality in Germany, best loved for his role alongside 1970s sex symbol Ingrid Steeger in the cult sketch-comedy show *Klimbim*. He was also famous as the actor who spoke the words of Robin Williams in German-dubbed versions of his films. Few Germans are aware, however, that he was also a multitalented musician, proficient on six instruments, including the cello, piano, and drums. But the events of 8 November 2005 put an end to his musical and theatrical career. That was the day Augustinski suffered a stroke.

An artery ruptured in the right hemisphere of his

brain. Blood flooded out, destroying a large number of brain cells. Sixty-six years old at the time, the entertainer was now paralysed on his left side, meaning he was no longer able to establish a functioning connection between planning and executing an action on the left side of his body. Although he was still able to imagine clenching his left fist, for example, his muscles would not obey the corresponding order from his brain because they no longer received those orders via his nerves.

## Neighbourhood help inside the skull

A third of stroke patients recover alone within a year, which is evidence enough of the enormous capacity of the brain to heal itself. Its usual 'strategy' is for the nerve cells in the immediate vicinity of those destroyed by the stroke to take over their function. However, they only do so if they are forced, so to speak.

Thus, if a stroke leaves someone only able to limp, the best thing is for the person to try to walk despite the reduced functionality, because that forces the cell structures surrounding the damaged area to jump into the breach for their destroyed colleagues. For activities such as walking, this often happens automatically because people will try to stand and walk despite their limp and other limitations and the difficulties these cause. At first, their attempts will be stilted and hesitant, but with time their walking becomes increasingly confident and wide-ranging until they may eventually almost regain the level of mobility they enjoyed before their stroke.

However, this is often not the case when patients'

hands and arms are affected. If the left side no longer works, patients still have their right hand and arm. This results in patients leaving their paralysed limb dangling uselessly while they concentrate on using their functional side. This means the cells surrounding the damaged area do not receive any training stimuli and the left hand remains permanently paralysed.

This effect can be countered if, soon after their stroke, patients are prevented from using their healthy hand. For example, by tying it up. Patients are then usually able to use their originally paralysed hand in daily life again within a month. This is contingent, however, on the training beginning while the extremity is still receiving sufficient residual signals from the brain. If that is no longer the case, such training is impossible, and, without training, there can be no change — and that was precisely what happened in the case of Peer Augustinski.

### Neurone gaps are not forever

When the actor came to us in 2007 — he had heard of our research with stroke patients from his doctor — his left hand was completely paralysed, and he had difficulty walking. His strength of will, by contrast, had not suffered. On the contrary, he was highly motivated and ready to take part in any experiments.

We trained him to use our magnetoencephalography machine, the MEG. This is a helium-cooled tank that sits on the subject's head like an old-fashioned hood-hairdryer. More than 200 sensors on the inside of the MEG helmet pick up the magnetic signals emitted by the brain as it works. A superconducting circuit transforms these signals into electric impulses that can be visualised on a computer screen (see illustration on page 109).

Augustinski sat beneath this encephalographic hairdryer. We were most interested in the signals from the part of his brain that had controlled his left hand before his stroke, so we asked him to move the fingers of that hand. However, we did not want him to just think about moving it, but rather to really send out a command *to* move it. Just as he had done so many times every day before his stroke. We wanted him to avoid thinking about the gap between the command centre in his brain and his left hand, which we provisionally bridged.

What that means is that we received the command signals from his brain via the MEG, we converted them into electronic signals, and if he activated the correct brain areas, we moved plastic clips that we had fastened to the fingers of his left hand. Augustinski had jokingly dubbed these clips 'thumbscrews', but, in fact, they were precisely the opposite: they made his fingers move just as they would if his brain had decided to move them — albeit with a time delay. So when he thought, 'I'll move my left index finger', his finger would be moved. Not as quickly as was the case before his stroke, but enough to convince his brain that the command it was sending out was having an effect. And when the brain sees that it can affect something, it eagerly deploys its enormous potential for learning.

Augustinski attended 20 training sessions with us. He was far from being cured, but, after this training, he was able to move the fingers of his left hand to some extent without the aid of the machine. The exact processes going on in his brain remained a mystery. We examined it with MRI and discovered a tract of nerve fibres that passed through the destroyed area of his brain. We were unable to ascertain whether it was a new development or had been there before and simply become thicker and therefore visible due to the training. It was also possible that it was completely unconnected to Augustinski's progress towards recovery. But it seemed as though his brain had managed to build a kind of neurological bypass.

Whatever the truth of it, his brain had learned that it could still achieve an effect, creating the necessary conditions for further training, which Augustinski completed with characteristic discipline and the aid of our highly committed physiotherapist, Doris Brötz. He regained the ability to use his left hand when driving a car, climbing stairs, or lifting a glass of water to his mouth. He also returned to the stage in September 2011, in the hospital-based farce *It Runs in the Family*.

## Training, not fading, aiming not flooding — what really benefits the brain

Strokes are among the great challenges faced by medicine. Some 15 million people are afflicted by them every year, and they are the second-most common cause of disability (after dementia). Despite this, progress in therapy for stroke victims has been limited because too much focus has been placed on pharmaceutical treatment.

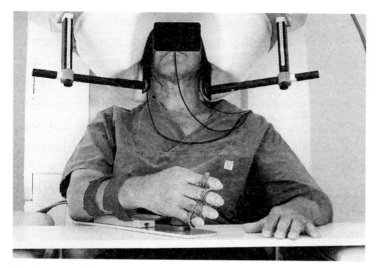

**Brain–machine interface (BMI) with magnetoencephalography device (MEG) for chronic stroke injuries**

The paralysed hand of the patient is attached to a neuroprosthetic (orthotic). When he thinks of moving his hand and suppresses the MEG's so-called sensorimotor rhythm in the brain, the orthotic device opens or closes his fist. This re-establishes the association in the brain between the command signal and the actual movement (the effect of the command signal). An electroencephalogram (EEG) BMI works in the same way, but with a less precise resolution. It is, however, more easily portable and cheaper.

In recent times, many victims have been treated with antidepressants, not only with the aim of helping them to come to terms better with their sudden disability, but also because such drugs are thought to improve the brain's capacity for recovery. However, research-based evidence for this is inconsistent. For instance, antidepressants often caused rather alarming side effects in studies with stroke patients. These included unsteadiness, disorientation, listlessness, and trembling, all of which sound more like an increase than a decrease in the symptoms of stroke survivors. Furthermore, an American study from the year 2009 even showed that post-menopausal women who were treated with antidepressant selective serotonin-reabsorption inhibitors (SSRIs) had a 45 per cent higher risk of suffering a stroke. Which begs the question of how a drug that increases the risk of developing a certain condition can then later be used to treat that very condition.

However, the pharmaceutical industry has other tricks up its sleeve. For example, there are currently great hopes for success with drugs that aim to use hormonal nerve-growth factors to promote the regeneration of damaged brain tissue. This may work in principle, but the problem remains that although new connections are certainly created in the brain, they are not necessarily *the* connections that would be helpful. And completely rewiring the brain in this way may lead to results as catastrophic as those achieved when it was believed that implanting dopamine cells was the answer to treating Parkinson's disease — those cells grew so rampantly that they caused tics so extreme that patients were almost unable to live with them. Flooding the brain with a drug is far more likely to result in a wholesale assault with various collateral damage than to strike the real enemy in a targeted way.

It makes much more sense to take advantage of the brain's enormous plasticity to help it heal itself. All this requires is to make sure the necessary impulses hit the right parts of the brain. When that is achieved, the brain is able to pull itself out of the swamp of its limited functionality by its own bootstraps, so to speak.

Thus, Peer Augustinski's progress was quite typical. We examined and treated 32 stroke patients, who had absolutely no residual movement in their arms and fingers, and who showed no positive response to therapy. We selected them from among 700 patients according to the following criteria: their hand had to be completely immobilised, they had to be neither depressed nor completely apathetic, they needed to have a person available to support and care for them during the treatment, and they had to live close by so that they could also practice at home. The majority were elderly patients. Training was successful for almost everyone, but of course more so for patients who still had an intact residual connection between their brain's command centre and their hands. After training, all patients were once again able to move their hands, and in many cases they were even able to use their hand in real-life situations (e.g. to clean their teeth, eat with a spoon, etc.).[10]

In future, we will feed the brainwaves from the cells of the brain's command centre through implanted electrodes directly to the prosthesis and the muscles so that each individual finger can be moved at will. A further advantage of BMI prosthetics is that they are portable and can be used by patients at home. This extends training times and makes it easier to apply the successes gained in training to the domestic environment.

I am hopeful that such BMI therapy will soon form a routine part of treatment regimes in rehabilitation centres and other institutions. However, psychologists currently show little inclination to include this method in their treatment repertoire. There is a great aversion to technology and biology in their profession, despite the fact their particular training means they should be the best practitioners to promote the effective learning processes that BMI therapy depends upon. This is not the place to discuss the reasons for that attitude. My hopes for the BMI rest rather with physiotherapists, sports therapists, and other non-academic professions working in rehabilitation.

## Even epilepsy is not an inescapable fate

Skull fractures, strokes, poisoning, encephalitis, tumours — the potential causes of epilepsy read like a litany of medical catastrophes, and that is why it is more common than we think: some 50 million people suffer from epilepsy worldwide. Sufferers face great psychological strain and anxieties. This is not only because they are literally thrown off course by their repeated seizures and must live with the constant fear of the next attack and the possible injuries it could result in; it is also because the people around them do not know how to react. Fyodor Dostoevsky was an epilepsy sufferer. He wrote, 'many cannot behold an epileptic fit without a feeling of mysterious terror and dread'. The same is just as true today as it was then.

And it is still true that any epileptic fit pulls the plug on the sufferer's brain. This is because the brain cannot metabolise the energy fast enough to cover the huge amount used during an epileptic episode, leading to the death of many cells by energy starvation. This can lead to long-term damage: the cognitive development of children with epilepsy is often delayed, and adults with epilepsy often suffer from cognitive decline. It was no coincidence that Dostoevsky called his intensely autobiographical novel about Prince Myshkin simply *The Idiot*.

All this is reason enough, therefore, to seek therapies to treat epilepsy. The truth is, however, that little progress is being made, just as there is little progress in changing public attitudes to the condition. The medical profession stubbornly persists in trusting in operations and medication, which turn out to be completely useless in one-third of patients.

Yet there are other, more effective procedures available for this condition, which also have fewer side effects. These procedures take advantage of the fact that neuronal activity registers most clearly as electrical signals during an epileptic seizure — which opens up huge possibilities based on the brain's powers of self-control.

One of my best friends and closest colleagues, the Italian neuroanatomist Valentino Braitenberg, described epilepsy as a 'short way from thought to fit', by which he meant that there is a continuum extending from normal excitability to the electronic salvos of an epileptic seizure. Dostoevsky described how, in the excitation phase, his brain, or that of his hero Prince Myshkin, 'suddenly amid the sadness, spiritual darkness and depression ... seemed to catch fire at brief moments ... I would feel the most complete harmony in myself and in the whole world and this feeling was so strong and sweet that for a few seconds of such bliss I would give ten or more years of my life, even my whole life perhaps.' The sense of awareness experienced by the otherwise rather melancholic author 'increased tenfold at such moments'.

Today, doctors describe this phase of a seizure as the 'aura'. During it, people with epilepsy like Dostoevsky can develop an almost euphoric burst of creativity. This explains why it really is just a short step from an innovative thought to a fit, which then however sweeps everything away with its destructive power. 'The face, especially the eyes, become terribly disfigured, convulsions seize the limbs,' Dostoevsky describes in his novel, 'a terrible cry breaks from the sufferer, a wail from which everything human seems to be blotted out.'

## Taking back self-control

The best way to imagine the aura phase prior to an epileptic seizure is to think of the neurons in the brain becoming increasingly charged with electricity until it all becomes too much and there is a massive discharge that overwhelms the person concerned in the form of a fit. In principle, it must be possible to prevent that final discharge from taking place, or at least to make it less severe by preventing the build-up of charge in the brain in the first place. In the past, many epilepsy patients managed this alone, developing strategies that enabled them to influence their aura, to suppress, dissipate, or neutralise it. This might involve clenching their fists, sniffing a bottle of perfume, pressing a finger against their upper lip, or trying to relax their mind by thinking of something that will return their brain to a calmer state.

It is thought that around 60 per cent of people with epilepsy could work with methods such as these and thus contribute to a mitigation of their seizures. The fact is, however, that only a very few make use of this potential these days. The psychoactive and anticonvulsant drugs that arose in the 1920s imposed a general policy of tranquilising the brain, meaning the aura phase was lost, and with it the chance for patients to prepare themselves for an impending seizure. The aim of this therapy is to stabilise sufferers, and this is indeed successful these days in two-thirds of cases. One-third are even left almost seizure-free. But such medication can also cause difficulty concentrating, slower reactions, and a general dulling of the brain.

This is a dubious strategy, and not only because it fails to achieve the desired therapeutic effect in a third of patients. It is also questionable because it places their fate in the hands of the medical and pharmaceutical industries, thereby exposing patients to a twofold loss of control: first, that caused by the overpowering violence of their seizures; and second, the removal of any opportunity to influence their condition themselves.

Even those patients who welcome such a pharmaceutical-based treatment because it brings relief from their fits must reckon with life-shortening side effects such as kidney and liver damage. For this reason, a learning therapy with no side effects would be a sensible strategy to complement drugs-based treatments, resulting in a reduction of the dosage of anticonvulsants needed by a patient without increasing the frequency of their seizures.

These considerations led us to adopt a 'back to the roots' approach. We wanted to give patients their aura back and then provide them with a few powerful techniques for effective preparation for an imminent attack with the aim of reducing its severity or, ideally, preventing it from happening altogether.

For obvious reasons, we selected the test group for our research from among patients who had not responded well to any drugs-based treatment. Some of them already showed unmistakable signs of cognitive impairment due to severe brain damage. We connected them up to the EEG and indicated to them via a computer screen what level of electrical excitability their brain was currently showing. For adult patients, this was represented by a coloured space rocket flying from left to right across the screen. If the rocket was green, everything was okay. If it turned red, this meant the patient's neurons were firing massively, indicating an imminent epileptic seizure. The subjects were instructed to try to prevent the seizure by controlling their brain activity and consciously switching the rocket's colour from red back to green. The aim: although seizures seldom occur in the laboratory, epileptics must learn to control the overstimulation of their brain, which is present even when they are not experiencing a seizure. That overstimulation is rendered visually on the screen as the EEG curve, an easily interpreted representational feedback for the patients.

Child test subjects with epilepsy watched a little man on the screen who was stranded on an alien planet and had to try to make his way to a nearby spaceship. The subjects could only make this happen by reducing the excitatory potentials in their brain — using nothing other than the power of their thoughts. And they had to do this as quickly as possible, since an aura can sometimes last only a matter of seconds. We treated mainly children aged eight or above whose parents and doctors has reached the conclusion that the side effects associated with drugs-based treatment were too damaging for the young patients' development.

Around a third of our patients with intractable seizures learned to calm their own brains. Age, intelligence, and gender had no influence on the results. Rather, the deciding factor was whether the patients had the necessary patience and stamina — that is, discipline — to persevere through the 30 to 50 hours of training required before a palpable effect appeared. One-third of our adult subjects even became totally seizure-free. For a chronic condition with no known cure such as epilepsy, this is a great success. The research involved 50 patients with severe epilepsy.

It was a difficult process for them because they also had to learn how to adjust their brains beyond the confines of our laboratory. To help them with this, we took our equipment into their homes at first. But ultimately, it had to work for them even when we were not there. Everywhere: while they were at work or eating out in a restaurant, while they were watching television or having sex, or, perhaps most importantly, when they were behind the wheel of a car. They had to incorporate the thought strategies and calming techniques they had developed into their everyday lives. Such a transference is not easy, and can soon become overwhelming, especially for a person with epilepsy whose brain is already severely damaged. Results showed, however, that one in three managed to restructure their way of thinking so as to prevent an imminent fit. A deterioration in cognitive abilities does not automatically mean the brain's self-correcting abilities are lost.

Despite this success for neurofeedback, which was reported in scientific journals specialising in research into epilepsy such as *Epilepsia*, this method is rarely, if ever, used, even in the treatment of the most severe cases of epilepsy, which would need it most urgently. There are many and varied reasons for this, and they are the same for almost all the learning therapies described in this book. One of the main reasons is a lack of knowledge of related branches of science. Although medicine and the health sciences have been paying lip service to the importance of an interdisciplinary approach for decades, it is still not practised. On the contrary, as the level of general knowledge increases, so does the degree of specialisation.

Doctors and neurologists are taught that physical conditions require physical — that is, medical (mostly pharmacological) — treatments. Psychologists are taught that psychological behavioural disorders require mental, and therefore social and psychological, treatment. This simple logic seems convincing, but it is wrong. Its main purpose is to artificially maintain traditional divisions between academic disciplines and professions. This in spite of the fact that many physical diseases of the nervous system, such as epilepsy, often respond better to psychological interventions, while mental and psychological disorders react better to medical treatment (e.g. medication to treat schizophrenia, or electroconvulsive therapy for depression). The lines separating physical and psychosocial causes are fluid and usually very difficult to draw.

Nevertheless, neurologists simply cannot imagine that such a severe brain disorder as epilepsy can be remedied through learning, just as psychologists are unable to conceive of using methods such as behaviour therapy and biofeedback to make a long-lasting positive impact on organic defects of the brain or body. Health insurers think in a similar way, as well as harbouring the legitimate fear of unnecessarily increasing costs by paying for pseudoscientific psycho-treatments. They therefore adopt the baseline policy of refusing to cover the cost for anything that does not fit with the conventional medical or psychotherapeutic way of thinking — and when drugs are considered to be an option, this inevitably also applies to neurofeedback.

A quarter of the people with epilepsy — those particularly severe cases who do not respond to treatment with drugs — are affected by this ignorance and withholding of treatment. Meanwhile, many extremely sick people, including many children, continue to take medication that is not only ineffective, but also associated with debilitating physical and mental side effects. This need not be the case, if only the individual therapeutic professions would look beyond their own backyard and concern themselves with the most important thing: what is best for their patients.

## Even half a brain will work

These days it is presumably common knowledge that we have inside our skull a brain that is divided into two hemispheres. While our left kidney has the same function as its counterpart on the right, and our left lung is responsible for breathing just as our right lung is, we are told that the two hemispheres of our brain are in charge of very different tasks. It is the general belief that language, logic, and rational thinking are located on the left, while creativity and emotions are found more on the right. It is now even possible to complete various tests to find out whether you are more left- or right-dominated — and if you are not satisfied with the result, you can engage in certain training procedures to change your hemispherical dominance. Many people prefer to think of themselves as right-brain dominated, since they see creativity and emotions as more desirable and attractive than logic and rational thinking.

However, this hemisphere model is now considered obsolete among brain researchers. Today, scientists believe rather that while the two hemispheres of the brain do have different ways of processing information along the lines just described, these merely reflect priorities and preferences rather than exclusive areas of responsibility. Non-linguistic functions especially are usually located in both hemispheres, and brain activity can be recorded on the right as well as the left side of the brain for language functions, too. In some cases, one hemisphere can even take over all the tasks of the other.

This was the case of one ten-year-old girl.[11] When the girl's parents took their daughter to the doctor at the age of three-and-a-half because of epileptic seizures, it was discovered that the right hemisphere of her brain was almost completely non-existent. The right side of her skull contained nothing but a space filled with more-or-less-functionless spinal fluid. Clearly, that side of her brain ceased to grow very early during her development in the womb.

Nonetheless, the little girl had developed just as any other child would. She was also able to see normally — which is astounding, because the loss of one hemisphere of the brain should lead to the loss of vision on the opposite side. Patients without a left hemisphere are blind in their right eye, and vice versa. But the brain does not always necessarily stick to those rules.

Magnetic-resonance imaging and various perception tests showed that the girl's left hemisphere had taken on practically all visual function. This demonstrates clearly that the brain is not made up simply of inflexible wired circuitry; it also contains a mechanism to organise the alignment of nerve cells with each other. This means that the type of work done by various parts of the brain is not static. The brain constantly monitors them against the tasks they have to complete and adjusts them as necessary, allowing areas of the brain to take on functions for which they were not originally intended.

In 2002, researchers reported the case of a seven-year-old Dutch girl whose left hemisphere was missing. For this reason, she had only limited control over the right side of her body and, unlike the little girl in the case described above, her field of vision was also incomplete. But the little girl in the Netherlands had a different skill: bilingualism. She was fluent in both Dutch and Turkish, despite the fact that the language centre of the brain is located in the left hemisphere, and so she shouldn't have been able to speak even one language properly, let alone two. Once again, it appears that the intact half of the brain completely took over the functions usually centred in the other hemisphere.

It should be noted that such feats of adaptability as these can only work if the structures of the brain allow it. This is of course more likely earlier in life than later; in the case of the little girl born without a right hemisphere, the realignment began in the womb. However, this does not mean that the plasticity of the brain cannot achieve amazing feats of adjustment even in advanced old age.

In 2007, the medical journal *The Lancet* published a report on a 44-year-old father who was admitted to hospital with mild weakness in his left leg.[12] Doctors examined his brain — and discovered a large black hole full of fluid rather than functioning neurones. Overall, the man had only about half as much brain mass as his average age peers. He had been diagnosed with hydrocephalus as a child, and that fluid in his brain had repeatedly pushed his brain matter against the inside of his skull and inhibited its growth in the process. Nonetheless, the man not only held down a job as a civil servant, but also had a wife and two children. After doctors drained the fluid and reinstated normal pressure levels inside his head, the man was able to walk normally again.

A similarly spectacular case, this time from the United States, is that of Terry Wallis. He became comatose after being involved in a car accident in 1984. He remained in a minimally conscious state, only able to give the occasional nod or grunt. Magnetic-resonance imaging of his brain offered no hope that he would ever regain consciousness. Yet after nine years, he began uttering words again. They were not connected up into meaningful phrases, but they were an astonishing development all the same. Doctors re-examined Terry's brain and found that new nerve fibres had formed in his cerebral cortex to bypass the non-functioning areas. In addition, his precuneus was active once again, and becoming more active all the time. This laterally located area of the cerebral cortex is pivotal in forming consciousness of both self and our surroundings. Its reactivation in Terry's brain was an indication that he was in the process of returning to waking life. And indeed, in 2003, after spending a total of 19 years in a coma, he was able to speak normally again and move his arms and legs.

From that point on, magnetic-resonance imaging showed even more clearly defined changes in his brain, in particular an apparent explosion of new nerve fibres in the cerebellum. Terry's motor skills continued to improve, although he has never managed to learn to walk properly again. Apart from that, he is now able to participate in normal life and social interaction just like anyone else. He has come to terms with the fact that he has no memory of the 19 years he spent in a coma, and with the fact that his wife now has three children with another man. It is possible that this mental resilience played a part in his freeing himself from what seemed like a hopeless situation. We do not know. Indeed, to this day, we also do not know what processes cause a patient to wake up after many years in a coma.

However, we are increasingly coming to understand that the brain is able to heal itself, even after suffering severe damage, by 're-inventing itself', as it were, in a process of restructuring and rewiring. Sometimes this is more successful than others, but the fact is that the brain's potential is far greater than we used to believe. Thus, we have no need to fear that our brain, and with it our sense of self, might suddenly disappear.

And anyway, the brain also has its own solution to the problem of fear.

# 6

# Brutal Confrontation

*controlling anxiety and depression
without medication*

Most spiders are perfectly harmless. Even most kinds of
the dreaded tarantula are safe, and those that are native to
Europe are not even able to pierce human skin with their
fangs. Only the bite of the European yellow sac spider
can sometimes be painful, but no more than a bee or wasp
sting, and, unlike those wounds, the bite of the yellow
sac spider has very rarely been known to cause allergic
reactions. Not to mention the fact that it is extremely
unusual to come across one of these rare, timid, nocturnal
spiders. To summarise: where I live, there is actually no
reason to be afraid of spiders. Cars, cigarettes, and even
electric toasters pose a far greater threat to Europeans
than do spiders.*

---

* Even in parts of the world with dangerous spiders, the danger of spiders
in general is greatly overstated, e.g. there's no need to fear the basically
harmless huntsman or much-maligned white-tailed spider in Australia.

Nevertheless, up to 10 per cent of people in Germany suffer from arachnophobia — fear of spiders. Nine out of ten of those people are women. But why does this phobia even exist, when spiders are so harmless? Some psychologists believe the more dissimilar an animal is to humans in appearance, the more common and the more extreme phobias of it will be. However, sticklebacks and tree frogs are hardly similar to human beings, and they don't provoke fear in anyone. Another explanation is based on the argument that it is the unpredictable way arachnids move that frightens people. Along with their apparent ability to appear suddenly as if from nowhere. But most arachnophobes will scream just from catching sight of a spider waiting motionless in its web beneath a bridge, in full view of any passers-by.

Is it because at some time during the evolution of *Homo sapiens*, an idea that any hard-bodied creepy-crawly is a potential source of danger was programmed into our 'primal knowledge'? After all, thousands of people die from scorpion bites every year in Africa and Asia. In fact, however, arachnophobia is much less common in those regions than Europe, where there are no scorpions at all. So why should evolution have allowed Europeans the luxury of a phobia of something that does not even exist here, and presumably never did throughout the history of human occupation of this geographical area? And why should evolution have 'blessed' almost exclusively women with that fear, although they are no more likely to encounter a dangerous arachnid than their male peers?

For us, a phobia of cars or nuclear power plants would

make much more sense, but if you show people from industrialised nations pictures of those objects, they tend not to show a strong emotional reaction: for example, there is no significant reaction in their thalamus — a very primal, from an evolutionary point of view, part of the brain, involved in relaying information about potential threats to the cerebrum. This lack of reaction is presumably because such modern potential dangers as motor vehicles and atomic power stations were not yet in existence when the brain evolved and because, especially in the case of automobiles, we have become used to them as a normal part of our daily lives and are desensitised to their dangers.

Yet this is not the case for spiders. Many people, even those who do not suffer from arachnophobia, will often react rapidly and violently to an object before they are even consciously aware of its presence. However, that is precisely why such a fear is as senseless as a compulsion to wash one's hands 30 times a day out of a fear of germs. And just as senseless as Pablo Picasso's habit of insisting that his family — including two restless young children — sit quietly together in a room for a minute's silence before any outing, because he feared something terrible might happen otherwise.

As understandable as it is, even post-traumatic stress disorder (PTSD) — following an accident, an experience of war, a severe illness, or a serious crime — is not rational. When someone is too afraid to ride a bicycle after an accident, or when a war veteran panics at the sound of sirens or aeroplane engines, we can understand their behaviour, but it does not serve any purpose for us or the person affected. If a big dog is bearing down on you and gnashing its teeth, it makes sense to run away in fear; if it is peacetime and you still scurry under a table in fear every time you hear a fire engine's siren go off — and you even know that your reaction is out of all proportion — you have a problem. And it is, in fact, a communication problem.

You can tell an arachnophobe as many times as you like that spiders are not dangerous, and she can repeat it to herself like a mantra until she's blue in the face — but that will not alleviate her fear. A concentration-camp survivor can assure himself a hundred times that he no longer faces torture and execution, but that will not stop him dreaming of it every night and panicking when he sees film footage from that time. The central problem with anxiety disorder is that sufferers — or more accurately, their conscious mind and conscious will — do not have access to the deeper areas of the brain where their fear originates (see illustration on page 129). That access is interrupted due to a weakening of the anatomical connections between brain cells. It's a breakdown of communication between neighbours, so to speak, who should be working together inside the skull, but who — for various reasons — have ceased to do so. However, that situation need not be a permanent one.

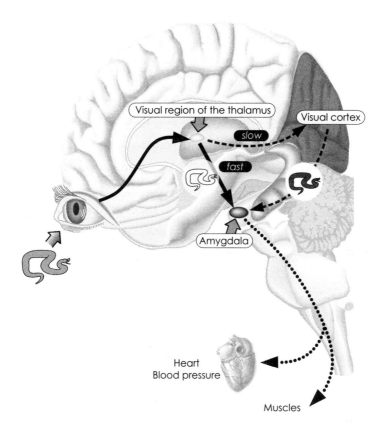

### *Development of fear, anxiety, and avoidance within the brain*

A simplified illustration of the nerve signal pathways from the eye to the brain and the body, using the example of seeing a dangerous snake. First, the visual information is relayed within a few milliseconds to the visual region of the thalamus (arrow from eye to brain), which triggers an initial, unconscious pre-processing of the perception and relays a rough copy of the perception 'snake' to the amygdala (bold black arrow), which activates the relevant parts of the body (muscles, cardiovascular system) (dotted arrows) for fight or flight. This physical stimulation is fed back to the cerebrum (not shown here, for the sake of simplicity), and, a short time later (approximately 200 milliseconds), the visual cortex (dashed arrow from the thalamus) and the attached memory centres are activated — and only then is the person aware of the fear and its cause (dashed arrow to the amygdala).

## Freud's idea of fear — and the false conclusions he drew from it

If I were to name the greatest common denominator in brain science — which is no easy task for such a complex organ — then it would be 'effect'. The brain *wants* to trigger effects, to initiate, to set things in motion, even without an ultimate aim. It wants to get rewards, pleasurable feelings; it wants dopamine and amphetamine. If the brain does not succeed in this, the areas that no longer achieve an effect and are no longer of any use are overwritten, and eventually wither away. Someone who had a couple of years of French at school but has never needed to use it will hardly be able to speak it in any useful sense at the age of 50. They might still be able to understand a word or two, but coming up with entire sentences will be a major challenge.

Our work with locked-in patients gave us a terribly haunting insight into the crucial role played in the workings of the brain by the expectation of effect. No longer having any access to their bodies, such patients are unable to achieve effects for themselves or anyone else, and that means their brains are in danger of descending into infinite indifference and lack of interest. Unless, that is, their brain can be convinced that it can still 'bring it', that it can still achieve effects. And, of course, some things are now possible that used to seem impossible, such as a totally paralysed man writing entire letters using only his own brainwaves.

In a way, anxiety disorders are the opposite problem. Take those people traumatised by war. Their panic response is totally disproportionate behaviour that does not achieve any meaningful effect. Quite the opposite, in fact: the war situation is over, and running screaming through your building will only cause disconcerted reactions in other people. Not to mention the fact that this behaviour renders the person concerned helpless and trapped by their own flight response.

That fear should be expected to disappear, since it achieves no sensible effect. But, in fact, the opposite happens, because there is a different, even more powerful effect involved! It may not be sensible, but it leads to positive consequences and thus a positive effect. The person concerned has no control over the senselessness of his or her panicked actions, but engaging in such flight and avoidance behaviour does bring relief from the deep-seated fear that's causing them. These people 'know' cognitively that their behaviour is irrational, can repeat that credo many times, and may even be ashamed each time they end up under a table in a state of panic. But that knowledge is isolated in the upper regions of the cerebrum and never reaches the lower areas of the brain, where fear reigns supreme. What remains is the immediate effect of reducing that fear by means of the rapid flight reaction, and that effect is enough to maintain the irrational fear response.

Added to this is the fact the fear response inhibits other parts of the brain. The connections between the over-trained fear centres and those brain regions that regulate fear reactions are strengthened, but only in one direction: from the fear system to the regulating system.

Connections going in the opposite directions atrophy and die away.

Even in his time, the Italian poet Dante highlighted in his work *The Divine Comedy* the powerful extent to which our fears are fed by our emotional unconscious mind. Sigmund Freud raised this insight to the central pillar of his theory of psychoanalysis. However, he drew the wrong therapeutic conclusion from it when he reduced the entire unconscious to memories of early sexual experiences or the parent–child relationship. And by maintaining that patients can be 'talked out of' their fears or trauma in psychotherapy — the success rate of this approach is extremely low, in particular for unconscious processes. It may happen that a phobia is eradicated during such a therapeutic conversation, but the usual effect is to strengthen it. It will only be weakened if the conversation leads to the emotional imagination deep in the brain being activated as violently as it was during the original traumatic experience, but without the much-feared effect occurring.

The brain needs to learn first-hand how it can overcome the fear in question without triggering a flight reaction, by directly experiencing the fact that there is nothing to flee from. So when a patient traumatised by experiences of war flees the psychotherapist's practice in a panic and learns immediately from that experience that there is no actual danger and he can safely return to the therapist's office, that experience — especially if it is repeated several times — may lead to the extremely well-ingrained fear mechanism in the brain being overwritten. The problem is that such extreme situations are difficult to bring about purely by talking. And when it happens, it usually happens by chance — and, as such, the phenomenon has nothing to do with systematic therapy or science.

Confrontation therapy, on the other hand, offers a prospect for treatment, which means 'confrontation with the real impossibility of flight'. This may sound absurd, since an impossibility is something that doesn't exist, but the meaning becomes clear on examination of a few examples.

## Confrontation rather than conversation

Fate sometimes works in convoluted ways to make an idea seem credible even though it is nothing but an astonishing cascade of coincidences. This was the case with Horst, a patient who later became a good friend of mine. He was a jeweller by profession, and nothing untoward ever happened to him in the exercising of that profession. But it was a different story when he was driving. Horst had no less than five car accidents in succession and they all followed the same pattern: while he was overtaking a truck, the larger vehicle's trailer suddenly swung out to the side and hit his car. After each collision, Horst's car was overturned. Not once, but five times in succession, within just a few years; on two occasions, his car even caught fire and burned out!

After the fifth accident, Horst threw himself out the window of the hospital he had been taken to. He later described this himself as an act of attempted suicide, and it had considerable consequences for his life: in addition to the brain damage from the accident, he also suffered brain damage from the fall, leaving him lying immobile in bed, stiff as a board, and unable to react in any way.

His doctor at the hospital diagnosed an organic brain syndrome with mutism. 'Mutism' describes a condition in which patients are able to perceive and comprehend their environment normally, but for some reason do not 'want' to react to it. Thus, this is different to the condition of locked-in patients, who want to communicate but *cannot*. The doctor's diagnosis of Horst's condition was certainly the most obvious one, since he had suffered various brain injuries and he barely uttered a sound.

Yet this medic was sceptical. Somehow, Horst did not resemble someone whose brain was irreparably damaged. Rather, he gave the impression of someone who had retreated into a kind of cocoon. The doctor researched his patient's background and discovered the incredible series of accidents he had suffered, which would easily have been enough to trigger a severe case of post-traumatic trauma and cause shock-induced paralysis. This meant the original diagnosis of irreparable brain damage might be wrong; indeed, the indications were that all was not lost for Horst. And so that scrupulous and self-questioning doctor referred his patient to my institute in Tübingen.

I did some more research into the jeweller's former life, and became convinced that his condition was due to trauma. He had not only experienced those five serious car accidents, but also — at that time of complete helplessness — been terribly hurt by his wife. She had realised that the death or disablement of her husband meant full possession of their house would pass to her, so she took advantage of his complete helplessness and confined him to the cellar. During one of the panic attacks that caused Horst to rattle wildly at the door of his basement prison and then slump weeping to the floor, she kicked him in the head with her stiletto-heeled shoe. Her aim was presumably either to kill him outright, or to injure him so badly that a legal battle between the two over possession of the house would no longer be possible on his part.

During the ensuing court case and therapy sessions, she described the stiletto incident as an act of self-defence during a marital dispute, although anatomical examination of Horst's injuries appeared to show otherwise. Eventually, possession of the house was awarded to Horst, but lawyers were not able prove a charge of attempted murder against his wife. Thus, he won a partial victory at court, but in the process his brain had suffered yet another injury, which presumably, in combination with his previous injuries, led to the severe memory disorders he suffers to this day.

Worse than those organic brain injuries, however, was the fact that the series of accidents he had suffered appeared to have left him with a — literally — paralysing fear, which we now needed to treat. We were aware that Horst had already tried almost everything, including psychotherapy, physiotherapy, and treatment with psychoactive drugs such as benzodiazepines (Valium), antidepressants, antipsychotics, and opiates. He had even tried homeopathic globules. But none of these had worked.

This was reason enough to try a completely different

strategy, based on a phenomenon described by Ivan Pavlov in the context of his research into the psychology of learning. This is extinction — the phenomenon in which a conditioned behaviour is prevented by means of repeated confrontation without the possibility of avoidance. This kind of therapy may seem shocking and even inhumane, but if it can save lives or return them to normal, it should be used — as long as it is used properly.

So, I strapped the motionless Horst into my Mercedes and drove off with him. We jumped red lights, overtook without indicating, executed emergency stops until the tyres squealed, and raced at breakneck speed down the autobahn — famously, the world's only major highway system with no speed limits. In short: I drove like a maniac — and the treatment began to take effect. My patient awoke from his lethargy and started screaming, whimpering, and crying. He vomited and evacuated the contents of his bowel and bladder onto my upholstery. Eventually, he slumped back in his seat, completely exhausted and unresponsive.

I repeated this procedure with Horst at least 30 times in the same way — just as his accidents had all occurred in the same way — but this time with *no* accidents: fasten seatbelts, drive recklessly, faeces, urine, vomit … and eventually calm. It may have been disgusting and unsavoury, but this is precisely what rendered the experience so intense that it caused an effective counterbalance to Horst's previous experience, reducing his accidents in the hierarchy of significance in his brain. He learned that our drives left him stinking dreadfully, but, otherwise, nothing bad happened to him: nothing. Everything was okay. No matter how wild I was behind the wheel. The only drawback was that this course of therapy left my Mercedes practically worthless on the second-hand car market!

Horst was not completely cured after this, but that was not to be expected considering everything he had been through. He was, however, once again able to communicate with those around him. One problem he continued to face was his recurring dreams about his car accidents, which meant those experiences were constantly being re-inscribed on his memory. He would often jump out of bed at night and run away, only waking up when he hit his head against the bedroom door. In his dreams, he always managed to escape the danger eventually, which was a positive effect that rewarded his fear reaction and therefore reinforced it.

It was once again Freud who first realised that any psychotherapeutic approach must also delve into the realm of dreams if it is to be effective. This is not — as Freud thought — because of the supposed symbolic meaning of night-time phantasies, but simply because dreams stabilise the embedding and interlinking of (possibly unwanted) memories. Horst continued to dream of his devastating accidents and was able to escape from danger successfully in his nightmares, making it difficult for the experiences he had with me to establish themselves in his brain as an alternative pattern (driving = fear = *no* catastrophic consequences).

The result of this was that Horst's phobia continued to return repeatedly, albeit often in a different manifestation.

There was a time when Horst was able to drive again, but was no longer willing to fly. When this occurred, I took him along with me on my long-haul flights, for example from Germany to America and back. I always carried a syringe of Valium with me just in case, but I never needed it: although Horst was anxious and peed his pants on more than one occasion, he survived the ordeal.

We would share a room in hotels, and he would sleep with me in my bed. At night, I would handcuff him to myself, so that he was unable to run away while dreaming. Sometimes, after one of his nightmares, he would jump out of bed, dragging me along with him like a man being drawn through the prairie behind a bolted horse in an old western film. Horst had agreed to this treatment, incidentally, because he was determined to get over his phobia. He was prepared to do and to suffer anything to be cured.

This constant confrontation — with the 'real impossibility of flight' caused by the fact that a determined psychologist was always at his side — enabled Horst to regain the ability to travel by car, ship, train, and plane. Today, he leads a mobile life, although he sometimes still cowers in fear in the passenger seat when he is in the car with me (though that might have more to do with my driving style). He leads a contented life and is able to spend time and laugh along with other people. And he is now separated from the wife who had bullied him so badly in the past.

Another patient, traumatised by his time in a Nazi concentration camp and longing to be free of his obsessive recurring memories, learned to live with them by, once again, being handcuffed to me — once again, with the patient's express permission — and being taken to visit a Holocaust museum, where we entered a reconstruction of the gas chambers. He relived his experiences, this time without them leading to pain and death. And this immediate emotional — not cognitive! — realisation was enough to free him from his tormenting dreams. It must be admitted, however, that the chances of successfully treating concentration camp survivors using confrontation therapy are limited. Nothing is truly comparable to the experiences of those who survived the industrialised murder machine of the Third Reich.

## Even dog shit can be a cure

Confrontation therapy can also be used to help people with obsessive-compulsive disorder, since this is also based on fear. For example, compulsive handwashing originates in a panic-stricken but rationally understandable fear of infection, and an obsession with control originates in a desire to protect against unpredictable accidents and disasters. The feeling of relief engendered by such obsessive behaviour is perceived as the reward for those fears and leads to their deep embedding and stabilisation in memory. They sit so deep, in fact, that a patient cannot receive a 'talking cure' from psychotherapy. A cure depends much more on teaching patients directly, by means of an intense, emotional confrontation, that they are not in any serious danger and that their fears are unfounded.

To achieve this, I and other therapists treated patients with a handwashing compulsion by taking them for a walk through Hyde Park (I had found sanctuary in London after I was thrown out of the University of Vienna for my rebellious behaviour there). We went round picking up dog dirt and smearing it on our faces. Each in our own face, it should be noted, since when confrontation therapists do not 'practise what they preach', they lose all credibility. In cases like this, the therapist must be seen to experience disgust, and to overcome that feeling as a model for the patient to learn to do the same. Patients were then instructed *not* to wash for a week. In real terms, this meant seven days of awful stench and social isolation, as no one wanted to stay around such a smell for long. No one, that is, except the therapist, since, when necessary, we handcuffed ourselves to our patients for control purposes. And, of course, that meant not washing ourselves either, otherwise the therapy would not have worked.

Relapses were extremely rare and could be cured by a renewed confrontation.

Confrontation therapy demands much more of the therapist than communication from chair to chair or from chair to couch. This is probably the real reason why psychotherapists these days prefer to try to treat their patients through talk. Not to mention the fact that today such a confrontation could lead to legal problems (legal liability guidelines have changed a lot) for therapists who handcuff themselves to their patients and then urge them to smear themselves with dog faeces or to jump a red light.

Various studies show, however, that confrontation

therapy is an effective treatment for anxiety and obsessive-compulsive disorder — more effective than psychoanalytical and psychotherapeutic talk therapy. And it has more chance of success than drugs-based therapies, which can inhibit the brain activity that is causing the problem, but cannot replace those activities with other, more desirable ones.

Confrontation therapy was developed as part of research into behavioural therapy and has been studied in many research projects. As mentioned earlier, it is based on the theory of extinction developed in learning psychology and neurobiology. Eric Kandel, who was honoured with a Nobel Prize in the year 2000, showed at the molecular level that repeated confrontation that does not result in the expected consequences leads to potassium molecules at the synapses (the connections between nerve cells) changing in such a way that the stimulation is reduced with each repetition. Pavlov, who received his Nobel Prize in 1904 and who described the optimum conditions for extinction, laid the foundations of behavioural and confrontation therapy. His laboratory dogs were clearly traumatised after nearly drowning when St Petersburg was hit by a flood. After Pavlov confronted them repeatedly with a simulation of the floodwaters breaking into the laboratory, they recovered and were able to take part in learning experiments again.

The aim of confrontation therapy is to force the brain to experience an intense and emotional 'light-bulb moment'. The brain must experience directly that the feared effect of a certain behaviour — and the emotional experience connected with it — *do not* occur, after all. This allows the brain to develop space in the memory for other things that do not have to do with the trauma or the obsession in question. An arachnophobe will still retain the memory of jumping on the conference table during a meeting at work because she saw a spider crawling across the carpet — but those memories will no longer have the previous paralysing effect. They leave the brain in enough peace for it to continue working normally, without being overwhelmed by negative emotions.

## Test jump from the Golden Gate Bridge

Charly was tired of life and wanted to end it all. He really wanted his suicide attempt to be successful. He didn't want to be one of those losers who make a song and dance of killing themselves only to fail miserably because they were not really trying to die. So Charly decided not to leave anything to chance. He got himself a pistol, a bottle of sleeping pills, and a rope — and with those tools in hand, he set off for the Golden Gate Bridge. His reasoning was that at least one of those methods of killing himself must work.

When Charly reached the bridge, he first took the sleeping pills. With red wine, to increase their toxic effect. Then he put the pistol in his mouth and pulled the trigger. However, his aim was wrong, presumably due to nerves, and the bullet missed his brain, leaving him with a bullet hole in his mouth. Charly pulled the trigger again and heard nothing but an empty click — he had been sold a gun with just one bullet in it. He tossed the gun into the water in anger and tied the rope to the bridge railings. His figured that if he failed to hang himself properly, he would still plummet more than 50 metres and meet his death when he hit the water. Or failing that, as a non-swimmer, he would drown miserably in the bay. But Charly tied the noose too loosely round his neck, and his body was slung feet first into the water. The sound his body made as it hit the water alerted the harbour police, who rescued him immediately. As he was recovered from the water, his stomach rebelled, perhaps due to the cold of the water or the stress of the situation, and he vomited — instantly neutralising the lethal effect of the overdose.

Since that experience, Charly's thoughts of suicide have never returned. On the contrary: he now works in social services, trying to convince people suffering from depression to abandon their thoughts of suicide. That is very much to his credit. But if he relies purely on the power of conversation to achieve that, he will not have much success. It would be far more effective if he put those suicidal people through the same ordeal he suffered at the Golden Gate Bridge. *Those* are the stimuli that really make a brain abandon any thoughts it might have of suicide.

Charly's case is one of the favourites in the rich treasure trove of psychological anecdotes. It makes us chuckle. Some people feel a kind of respect for the unlucky suicide-attempter's dogged determination to try *everything* he could to kill himself. What we see is that he went to the Golden Gate Bridge with the intention of committing suicide. The emotion centres of his brain were fixated on the idea that he was about to end it all. His pulse would have been racing, his veins throbbing in his temples, his body trembling, his thoughts revolving only around his approaching end, which would be final and forever. Yet in the end, it was not his final end that came, but rather — nothing. Thus, the desired effect never occurred. There is no better way to convince the brain that suicide is a completely unnecessary act.

Of course, there is the argument that men especially who are prevented from attempting suicide (women attempting suicide are less likely to succeed then men) will not infrequently try again, and are often then 'successful'. However, such repeated attempts usually only occur if the first attempt was interrupted at an early stage in the chain of behaviour, with most of the steps towards an actual attempt at suicide never completed. In such cases, the person attempting suicide does not experience the *entire* chain of behaviour (preparation and failed implementation) as having no effect, but only the beginning of the reaction chain. If, by contrast, the action is carried out to completion (such as when the gun goes off but 'only' causes a non-fatal injury), the recurrence rate is considerably lower.

In a therapy situation, it is difficult to reconstruct a suicide scenario such as that experienced by Charly. It would involve taking the patient to a bridge, encouraging him to jump and then holding him back at the very last moment. For a period of time, we actually did that with our patients, with therapeutic success. When patients expressed a desire to kill themselves, they were asked to 'walk through' the entire chain of behaviour with a therapist, right up until just before the actual act of suicide, and then imagine that act as vividly as possible. However, there is a danger that this can become nothing more than training for the actual act if the expected negative side effects experienced by the patient are not sufficiently intensive or physically negative. This approach should only be attempted by experienced therapists.

Still, our results indicate that depression could also be treated using confrontation therapy. This could be done not only by showing patients that their suicidal, everything-is-pointless thoughts do not result in a desired effect, but also by showing them that they are not in fact as powerless and helpless as they feel.

The film *Fearless* (1993) shows what such a confrontation might look like. In this movie, a young woman called Carla (Rosie Perez) becomes depressed because she blames herself for not holding on to her baby tightly enough during a plane crash in which the child died. Those feelings of guilt lie like a lead weight on her soul, and no therapist has been able to help her. She then meets Max (Jeff Bridges), a fellow survivor of the same plane crash. He tries to help the Carla by taking her to buy Christmas gifts for her perished loved ones. However, that doesn't help — because giving presents to dead people does not achieve an effect.

Max then straps his depressed friend into the backseat of his car and has her hug a toolbox as tightly as she can. He tells her to imagine she is back on the doomed plane, holding her son in her arms. Then Max crashes the car head-on into a wall. Both survive, albeit with serious injuries. Although Carla was prepared for the impact, she was still unable to hold on to the toolbox. She learns from this experience that she never had any chance of keeping hold of her baby during the plane crash. The result was that her brain immediately realised that her feelings of guilt were baseless.

The brain mechanism involved here is the same as that involved in the process of extinction as described earlier. Carla gets her life back. She leaves her husband, who is only interested in how much the insurance payout for their son's death will be. And she also leaves Max. He may have cured her of her depression, but he is also a physical reminder of the sad times she now wants to leave behind her.

## Moving forwards out of the backwards of depression

It must be said that there are many differences between films and computer-generated virtual realities on the one hand and actual reality on the other, and that depression rarely disappears as result of one single action. Those who have battled against depression for years cannot be cured in one fell swoop. Once a certain thought and behaviour pattern has become fixed in the brain, it takes a corresponding effort to dislodge it. Yet there's no reason to view depression as an irreversible fate, which can at best only be attenuated somewhat with pharmaceuticals and electric shock treatments. There is something that can be done, but it takes some effort and is based on the specific way our brains work.

A certain area of the frontal brain, below the corpus callosum, plays an important part in depression: Brodmann area 25, where a large number of brain functions converge. These include our sleeping–waking cycle, which is in itself important in depression, as the condition is almost always accompanied by insomnia. This area of the brain is also a significant driver and director of the brain's 'thought pump'. When Brodmann area 25 is highly active, our attention is focused, and so it is necessary for concentration. If there is less activity in this area, on the other hand, we are easily distracted, which can be difficult if we need to concentrate, but can also be extremely relaxing.

In people with depression, Brodmann area 25 is significantly *over*active. We still do not know the precise reasons for this hyperactivity, but scientists believe that it is triggered by experiences of loss. Many patients with depression have suffered a drastic experience of loss, especially during puberty, such as parental divorce, the death of a loved one, or a move to another city or even another country. They don't necessarily reveal this in talk therapy, but it is almost always present in their personal history. We believe this experience of loss then over-activates Brodmann area 25, causing the patient's thoughts to return again and again to the idea that there is no point to anything because everything always goes wrong anyway. Just as the patient's parents' marriage went wrong, or their beloved original home was suddenly replaced by a new, unwanted dwelling. Helplessness and lack of control unite!

Any therapy must therefore target this thought pump with the aim of weakening its effect or redirecting it. One way to do this is by using electric shocks to curb its activity, which has yielded impressive results in the treatment of depression in some studies. However, this approach is also associated with side effects, including lack of concentration and learning problems, because it also switches off Brodmann area 25's function in focusing attention. This occurs because the electric shock triggers an epileptic seizure, which, like a blow to the head, disperses memory for a time. This can become a problem if the patient's condition recurs and electroconvulsive therapy has to be administered repeatedly, potentially leading to permanent memory loss. Nevertheless, severely depressed patients usually agree to this serious intervention because their suffering is so great.

Cognitive therapy, however, has fewer side effects with a similarly high level of treatment success. Unlike conventional psychotherapeutic approaches, it does not seek to wrap patients in cottonwool, but confronts them head-on with their problem. For example, when a patient says he no longer wants to go to work because he is always being put down by his boss, a skilled therapist will immediately intervene and steer the conversation towards the many times the patient *did not* have a problem with his boss and perhaps even received praise from her. In such cases, mere talking can help tackle the problem. But this should optimally take place *in motion*. Thus not in the classic scenario, with the therapist and the patient sitting facing each other, or with the patient lying on the infamous couch, but rather with the patient moving forward as the conversation takes place. This evokes for the patient the opposite of his static, nothing-to-be-done attitude caused by his depressive state. It gives his brain the feeling of moving forward again at last after all the numbness and backward-looking ('Why should I ever succeed at anything in the future when I've never succeeded at anything in the past … ?'), and everything the patient learns during this process will be stored in his memory in the context of this forwards motion. Not to mention the fact that walking and running increase the production of dopamine and other neurotransmitters, which leads to an increase in interest generally and an improvement in mood. It takes about 15–20 hours of cognitive and movement therapy for an improvement to set in.

Despite this, there are many areas of day-to-day life in which a therapeutic conversation, even in motion, is not enough to break the avoidance patterns typical of depression. For patients who believe they have no chance of finding a partner, it is of little help to reassure them that 'there's someone out there for everyone'. If they were encouraged by such a platitude and actually joined an online dating agency, for instance, and were then rejected by eight out of ten possible partners, they would only see this as confirmation of their original opinion that it is useless to even try to find a partner. Even though eight rejections from ten attempts is not such a bad average, as it means that two of those ten were successful. But a rational view is precisely the one that depressed people will *not* take, and no amount of persuasion will convince them. The solution then is to arrange contacts for such people (which can work even if the other person receives money for the arrangement).

A person who shuts herself away in a darkened apartment all day needs to go out into the sunshine. Talk is not enough.

At first sight, it might seems strange for therapists and their patients to embark on various social and sometimes risky escapades together. But how else is the brain to learn that the actions it has been avoiding do not entail the consequences it fears? That, on the contrary, these actions can achieve a desirable effect? Well-meaning conversation is usually not the way for it to learn this lesson — anyone who hopes otherwise might as well hope that lifting weights with just one arm will lead to the other arm gaining muscle mass, too.

Meanwhile, there is still no hard evidence that antidepressant drugs are any more effective than placebos. Marcia Angell, the first female editor-in-chief of the world's most prestigious medical journal, *The New England Journal of Medicine*, undertook a thorough comparative study of all the scientific literature, including that published in her own journal, on the subject of antidepressants.[13] She came to the devastating conclusion that most studies are so badly designed or distorted by the pharmaceutical industry that any interpretation of them is useless. The few methodologically sound research projects showed that antidepressants have scarcely more effect than placebo tablets. And although lithium compounds were found to offer effective treatment of manic-depressive disorders, their side effects are often very severe.

Although it no doubt requires great commitment from the therapist, and sometimes comes into conflict with traditional concepts of morality and even the law, confrontation therapy enables us to achieve the necessary intensity required to make use of the brain's plasticity in the treatment of phobias and depression. No other approach offers such prospects. So, before agreeing to pay for psychotherapy, health insurers and authorities should promote real, targeted confrontation therapy, in particular for the treatment of phobias, depression, and addiction disorders.

But what is the solution if the problem is not anxiety or depression, but precisely the opposite: a total lack of fear and remorse?

# 7

# Lock Them Up and Throw Away the Key?

*even psychopaths can change*

Stefan was successful, charming, amusing, sometimes quick-tempered, but otherwise friendly — a perfectly well-regarded member of society. Then, together with a friend of his, he was accused of kidnapping several women and abusing them. Could such a man really be a ruthless and cold-hearted criminal?

His case was heard in court in the summer of 2008, and it left the public, and even hardnosed lawyers and journalists, aghast. Not only because of the crimes themselves — videos of some of which were able to be shown in court, because the two men had filmed them — but also because the 42-year-old web designer displayed absolutely no emotion during the entire trial. He was obviously totally indifferent to the suffering of his victims.

The story of his crimes begins in August 2006, when he

and a friend placed a contact ad in a newspaper offering 'attractive and well-paid part-time jobs'. When a foreign psychology student replied, Stefan and his friend asked her to meet them at a two-storey house, from which she was not to re-emerge for three months. The 23-year-old woman was stripped naked and wrapped in cling film, kept in a cage and raped repeatedly, and eventually forced into prostitution. She was made to eat her food, mostly dry pasta, from a dog's bowl and was led round the house on a dog's leash by the two men. A few days later, she was joined in captivity by another woman, who suffered similar abuse at the hands of the two men. The second woman was also forced to have sex with paying customers, all of which was filmed using a camera hidden in a teddy bear. The two victims' ordeal did not come to an end until the criminals kidnapped a third woman, who managed to escape through a skylight — naked, with her hands still cuffed together.

During the court case, Stefan's accomplice showed remorse for his crimes, and it was generally believed that his feelings were genuine. This did not result in any leniency from the court or sympathy among the public, but people were willing to accept that there was some spark of empathy in him. The same was not necessarily true of Stefan. He showed no emotion at all, and certainly uttered not a single word of remorse. His strategy was a different one: he hoped to be spared jail by arguing that he should be classified as mentally ill. He believed that it would be easier for him to influence things in his interest in closed psychiatric care than ordinary prison — although he did not give the impression that he was in any way afraid of a lengthy prison sentence. Stefan's attitude rather resembled that of a Monopoly player trying persuade a fellow player to sell him Mayfair. But his plan backfired.

A psychiatric expert confirmed sadistic and narcissistic personality traits, as well as an antisocial personality disorder, leading the court to sentence Stefan to 14 years in jail with subsequent preventive detention, due to the 'threat he posed to the community'. This was a tough sentence, but for many it did not go far enough.

The widespread opinion is that such psychopaths should be locked up forever. Or even sentenced to death in cases like that of the American serial killer Ted Bundy or the Norwegian terrorist Anders Breivik, because their brains cannot be cured. After all, no one would have thought of placing Adolf Hitler in the care of a mental institution. The logic here — which is certainly not confined to the pub — is that of the saying 'better an end with terror than terror without end'. However, this argument is doubly deluded: it fails to recognise the typical characteristics of a psychopath; and it fails to take into account the plastic nature of the brain.

Hitler — like most politically motivated mass murderers — was far from being a psychopath. For one thing, he was too asexual to be a psychopath; for another, he was too emotional and unstable. Psychopaths do not cling on to their plans, come what may, but abandon them without a second thought when the plans stop working the way they would like and fail to deliver the intended results. Hitler, by contrast, persisted in his delusional pursuit of ultimate victory well after everything was lost. If he had been a psychopath, he may have been able to be stopped at an earlier point — because what is true of all of us is also true of psychopaths: their brains can be shaped and reshaped.

## From the prison cell to the business lounge

'If crime is the job description, the psychopath is the perfect applicant,' wrote the Canadian psychologist Robert Hare. This is because psychopaths lack the inhibitions that prevent a healthy person from wilfully lying and breaking the law, from torturing and killing. More than half of all serious crimes are committed by psychopaths. However, this does not mean that psychopaths are to be found only in prison cells, since they are often clever enough to avoid being recognised as such. Not to mention the fact that the typical characteristics of a psychopath, such as brutality, fearlessness, adventurousness, sensationalism, overconfidence, and lack of empathy, are conducive to a rapid rise up the career ladder in most societies.

This is why a great number of psychopaths can be found in the upper echelons of multinational corporations, churches, the military, universities, political parties, and other institutions. And if a successful surgeon is able to perform operations without showing a hint of nerves, coolly making incision after incision without ever hesitating or doubting his competence, it may be because he has managed to train himself over the years not to think of his insecurities or fear of failure — or it may indicate that he was a psychopath before the start of his career, who managed to find a niche where the symptoms of his disorder could develop in a socially respectable way.

Hare estimates that about 4 per cent of people in the business world (managers, executives, agents, and traders) display extreme psychopathic traits. He estimates the overall proportion of psychopaths in industrialised societies to be between 1 and 2 per cent. The figure is often thought to be higher among men than women, although this is doubtful. The fact that there appear to be fewer female psychopaths probably has less to do with a less-pronounced propensity for psychopathy, and more to do with the fact that overt psychopathic behaviour is tolerated and rewarded less in women than in men. This does not, however, mean that female psychopaths are more easily conditioned than males; rather it means that just like their male counterparts, they are skilled at adapting their behaviour to fit in with the prevailing conditions in society. Scientific research has shown that women are just as prone to psychopathy as men, but their brutality is more likely to be expressed psychologically than physically, and they have less of a tendency to commit sex crimes.

It seems safe to say that the majority of psychopaths do break the law, but they do not by any means all end up as criminals. Many lie and cheat, exploit others, and accept no responsibility for their own actions. As relationship partners, they are often unfaithful; as bosses, they have no care for the future of their company or employees. Their ability to manipulate people is unparalleled.

In early 2009, Canadian legal psychologists examined which prison inmates manage to gain early release from their sentences. Parole boards would have every reason to keep psychopaths under lock and key, since they traditionally have very high rates of re-offending. Despite this, psychopaths were found to be almost three times more likely than other inmates to persuade experts that they no longer posed no threat. They played the role of the remorseful sinner convincingly and said exactly what was expected of them. Even scientists are not immune to the cunning of psychopaths.

## A short test for psychopathy

Do you have psychopathic tendencies? The following test (based on Robert Hare's PCL-R2 psychopathy checklist) can help you find out. Simply tick the box that corresponds to your answer to each question. Of course, the test will only work if you answer the questions honestly. Please note: lying on a test like this — for example, because you realise a certain answer will push you closer towards psychopathy — is itself an indication of psychopathic tendencies.

| Question | Fully applies to me | Somewhat applies to me | Doesn't apply to me |
|---|---|---|---|
| I will do almost anything to get attention. | | | |
| I can be very self-assured, opinionated, even a braggart. | | | |
| Even early in life, I repeatedly cheated, conned, or defrauded others. | | | |
| I often don't feel much concern for possible losses or suffering of the persons I leave behind. I feel that some of them didn't deserve any better. | | | |

| | | | |
|---|---|---|---|
| If something goes wrong, I rarely take full responsibility but tend to manipulate or to put the blame on others to divert attention away from my own failures. | | | |
| I had sexual experiences with others at a very early age (12–13 years). | | | |
| It is hard for me to commit to long-term relationships. Sexually, I tend to be unfaithful to my partners. | | | |
| I get bored easily. I repeatedly feel a strong need for novel, thrilling, and exciting stimulation. | | | |
| With 'classic' forms of work, I often feel a lack of motivation. I hate routines and need variety. | | | |
| I like to gamble and often try to get hold of money by gambling or from other risky projects. | | | |
| I tend to be cruel to animals. | | | |
| In my life, I have done many things that could be considered criminal by a court, even if I have never been arrested or convicted so far. | | | |
| I do not shy away from physical or verbal aggression, which I see as legitimate means to get what I want. | | | |
| I have a good memory, which I mainly use to remember people's weaknesses. | | | |

If you ticked 'fully applies to me' more than eight times, you should undergo a full test for psychopathy carried out by an expert.

## The psychopath and fearlessness syndrome

How do people become psychopaths? Are they born or made? Researching the family backgrounds of psychopaths often reveals factors that are conducive to mental disorders and behavioural problems, including alcoholism, violence, poverty, and neglect. But in about the same number of cases, none of these factors are present. Children may grow up in completely normal middle-class families, in an atmosphere of respect, discipline, and family warmth, and still develop psychopathic characteristics. In such cases, it appears to make sense to consider genetic factors as the main cause and, as genes cannot easily be changed, psychopaths are therefore generally considered impossible to treat. However, there is also a tendency to cite psychopaths' family backgrounds as the reason for their resistance to therapy. The thinking goes something like this: children who grow up in difficult family circumstances learn such behaviour patterns early in life and they are embedded so deeply in the brain that they can no longer be changed. But whether genes or family circumstances are to blame, there is widespread agreement that psychopaths cannot be cured of their disorder. This consensus, however, is based on a lack of knowledge.

The fact is that psychopathic behaviour does not originate from an abstract, amoral mental attitude or a particular personality trait, but from the failure of certain interconnected areas of the brain to function properly. The four most important regions involved are the following (see illustration on page 32):

1. The amygdalae are two almond-shaped areas in the brain. They are the processing centres for emotional stimuli from the environment and for such stimuli's effect on autonomic functions such as breathing,

heartrate, and muscle tension. The amygdalae are considered to be the brain structures responsible for the *emotional response to information*: they react predominantly to fight-or-flight situations, and that reaction is usually very rapid, taking place even before we are conscious of it. The amygdalae of psychopaths are smaller and less-well supplied with blood than those of others.

2. The cingulate gyrus. It is located above the corpus callosum, which connects the two hemispheres of the brain. It influences *pain processing, regulation of emotional responses*, and *awareness*. Monkeys whose cingulate gyrus has been removed are tame, but also lose all interest in the other members of their social group. This area is also involved in the *long-term storing of negative memories*. Its reduced function in psychopaths means, among other things, that they are less able to remember negative encounters with other people and therefore less likely to learn from aggressive confrontations.

3. The insular cortex. In evolutionary terms, this is an ancient part of the cerebral cortex. It is located just above the temporal lobes, and one of its functions is to inform us of our emotional state on the basis of autonomic changes such as an increase in heartrate or muscle tension. Doctors call this the organ of *interoceptive awareness*. The anterior part of the insular cortex is also involved in empathy: neuroscientists have shown that it is more active when we empathise with other people's suffering. Such an increase in

activity could not be found in the insular cortex of psychopaths, who are completely indifferent to other people's suffering.

4. The prefrontal cortex. This covers many areas with many different functions, including telling us what is likely to be good or bad for us in the future. *After* we act, it produces feelings of guilt, regret, and remorse, but also emotions such as triumph and joy. You could call it the 'Biblical part' of our brain. In psychopaths, it only functions in one direction: psychopaths know what is good for them personally, but cannot judge what is bad for others and even for themselves. This is why psychopaths often act both impulsively and ruthlessly.

Psychopathy is fed by the inaction or reduced functioning of various areas of the brain, which, however, all have one thing in common: they are all involved in some way in our fear response. The American psychologist James Blair showed that although children with psychotic tendencies are very desirous of rewards, they are unaffected by punishment. No amount of telling off or knuckle rapping will affect them. No matter how many times it is explained to them that certain behaviour is damaging to their group and to themselves, they will continue to engage in that behaviour if they believe it will benefit them personally. And this pattern continues in adult life, since all punishment or the threat of it is ineffectual. While in other people sanctions lead to the development of fear and avoidance behaviour — often the appropriate conditioning of a model is sufficient to achieve this — such conditioning does not take place in psychopaths. They continue with their behaviour, completely unaffected by the anticipated negative consequences of their actions.

The brains of psychopaths know neither fear nor anxiety. Cognitively, they know the consequences of their behaviour. They are well aware that a person will die an agonising death if they are locked in a tiny room with no air supply, and they are aware that they will face severe punishment if they are found out. But that knowledge does not trigger any negative feelings in them. This indifference is sustained by the reduced functioning of several areas of the brain, and by the silencing of a complex neuronal functional circuit that normally slams on the brakes if there is a threat of punishment, crisis, or disaster, by signalling to the brain's owner that a given behaviour is likely to lead to fearful consequences. Of course, all this gives little cause to hope that psychopaths can learn to behave morally.

However, there is more encouraging news: the functional circuits for fear in particular are an extremely plastic part of the brain. From an evolutionary point of view, it would make little sense for them to be immutable in an ever-changing environment. Rather, those parts of the brain must be especially flexible, since all living beings need to be able to avoid danger if they are to continue to exist; and since the environment constantly poses new dangers, the areas of the brain involved in fear responses, which are so necessary for that avoidance, must be capable of adapting and changing in response to changing dangers.

This is also the case for psychopaths. Although the fear centres of their brains function only weakly, or may even be underdeveloped, they can certainly be stimulated, encouraged, and further developed. It is difficult to predict how far that process could eventually go, and this is something that must be investigated in empirical studies. But we know that it *does* work, and so may harbour great opportunities for treatment.

## Learning to fear again

In the early days of behavioural therapy, electric shocks were used to treat psychopaths. Sex criminals would be shown various pornographic films, some of which were violent and abusive. If the subject developed even the first indications of an erection while watching scenes of violence, a measuring device would detect that tumescence, and transfer the information to a computer that would trigger a severe electric shock. Nothing that would cause real harm, but enough for the brain to associate arousal at watching a violent sex scene with an intensely negative feeling. Arousal while watching non-violent pornography, on the other hand, was unsanctioned and sometimes even rewarded — for example, by allowing the subject to continue watching and by even providing him with snacks, just like a comfy evening in front of the TV.

Treatment with electric shocks was very successful because it perfectly targets the brain's way of working. Psychopaths don't lack the ability to feel pain, just the normal negative expectation of it, and so through the frequent and extreme association of pain with the images, they were able to learn to avoid that punishment stimulus by suppressing their positive sexual reactions to it. The electric shocks employed were slight, far below a level which could damage tissue, and also too slight to produce any trauma — and certainly not any additional aggression, since they were administered by a nameless computer so that the patients were not able to identify them with any actual person. It was simply that any incipient erection triggered by sexual violence resulted in an extremely unpleasant stimulus, while a sexual reaction to non-violent scenes had no negative consequences, without any reference to a particular punisher or non-punisher.

Despite this success, electric-shock therapy failed to gain favour, and indeed was discouraged as a brutal and inhumane treatment method. It was too reminiscent of Stanley Kubrick's film *A Clockwork Orange*, in which a sex criminal is administered a drug to make him sick when he's forced to watch violent porn films. Many cinemagoers looked away during that scene and even felt sympathy for the criminal. It is remarkable that a society that thinks nothing of locking psychopaths away for life rejects this effective treatment because it is seen — falsely, from an objective point of view — as inhumane.

Of course, there are fundamental ethical concerns about aversive learning, even when fully informed subjects have agreed to undergo the procedures. Those concerns rest on the apparent similarity with education or parenting — contexts in which the physical punishment of children is forbidden. The only problem with such a comparison is that patients with mental disorders are usually adults who are willing and able to give their consent. Furthermore, in a therapy situation, electric stimuli caused far fewer unwanted side effects than verbal punishment. Research has clearly disproved any 'transference effects' or 'displacements' in the Freudian sense, in which aggression is simply suppressed, only to surface elsewhere all the more violently after the aversion therapy is ended.

What remains, of course, is the critical argument that such a procedure is an affront to 'human dignity'. Although such an argument is defined in vague terms and difficult to disprove, it must be taken seriously. Before the start of any such treatment, there must be careful consideration together with all those involved in the aversion therapy (which also includes victims and potential victims in the case of criminal subjects!) of the possibility and extent of any insult to human dignity or danger to physical or mental integrity. In most countries, this takes place under the auspices of an ethics committee.

However, other, softer treatments also exist. They aim to make the physiological responsiveness of psychopaths more sensitive as a way of reanimating their affectivity. This can be achieved using biofeedback to train subjects to develop a feel for autonomic changes such as an increase in their pulse rate and sweat production or fear potentials in the brain. To do this, patients may follow an easily understandable representation on a computer screen of the changes in those physiological signals while they imagine something that triggers an emotion. Feedback gradually enables patients to learn to regulate those physiological signals and also to increase the intensity of formerly weak body signals, just as if they were learning the physical movements needed to play a new sport. Anybody can do this, although psychopaths require more time.

Autonomic reactions, however, are ultimately processes that don't necessarily reflect fear-related activity in the brain. So we in Tübingen have taken this concept a step further, namely directly into the brain itself, where we can use neurofeedback instead of biofeedback.

Our procedure involves placing the patient in a magnetic-resonance scanner (see illustration on page 26), which analyses the activity (blood flow) inside the brain in real time, focusing on the fear centres described above, such as the insular cortex. The neuronal activities in the fear centres of the brain are represented as an image on a computer screen, for example as a thermometer. It shows a rise in temperature when the target areas of the brain are particularly active, even though the patient feels nothing. The patient's task is to try to 'raise the temperature' at will. *How* this is done is left entirely up to the patient: using emotionally charged memories or visualisations, with abstract thoughts, or perhaps by thinking of nothing.

Even the fear centres of psychopaths receive a minimal amount of blood flow, and tiny changes in that can serve as the base for the learning process, so that the patient gradually learns how to influence the thermometer and 'raise the temperature'. Once the blood flow to the selected areas of the brain has reached a certain level, the test subject will begin to experience feelings of fearful anticipation. The aim is then to reinforce those feelings, which usually succeeds, despite the negative nature of those feelings of fear, because the brain receives immediate feedback when the increase in activity is achieved. To put it another way: the brain experiences a direct effect, and this is extremely motivating for the brain.

Healthy individuals usually learn to control the thermometer within two or three training sessions. Psychopaths require more time — 12 to 16 sessions — but they also eventually manage to activate their stunted, barely functioning fear centres. In some cases, brain scans also showed an increase in the size of the patient's amygdalae. Post-training interviews of psychopathic patients showed they were less likely to be thinking of general horrors such as car crashes or concentration camps, and more likely to imagine events from their personal lives, such as the death of their grandmother or their last court sentencing. But that is basically irrelevant. What is important is the fact that psychopaths can learn to reanimate the underdeveloped areas of their brain using neurofeedback techniques.

That activation means there is also a change in their perception and emotional responses. We were able to show that this is the case using an experiment in which we showed our psychopathic patients grisly images, including those of starving children or maimed war victims. Before their neurofeedback training, the patients were relatively unfazed by those scenes; after treatment, their reactions were significantly more sensitive, as could be shown by an increase in the activity of their sweat glands, as measured by means of skin resistance.

This reanimation of the brain's fear centres even

enabled us to positively influence psychopathic violent criminals who were deemed totally resistant to any corrective influence. We had one patient who had got into trouble even at school for drug-taking, and as a violent, cold-blooded thug with extremely weak academic achievements. His father was an old Nazi determined to turn his son into a 'fine upstanding German', in part by beating him regularly. The patient's sister was also abused, while her brother looked on impassively. The father's death was a release, and our patient made no secret of it, holding a big party at the funeral. He went down the path of delinquency, working his way up to a high position in criminal circles, where his unwavering, emotionless brutality earned him respect. He also sometimes worked as a bouncer at various bars and brothels, and as a 'debt collector', known for his 'heavy hand'. However, his violence began to get out of control. Eventually, he started beating up people who had done nothing to annoy him and who he had not been employed to beat, simply because they happened to be nearby. There was no rhyme or reason to his choice of victims; he was completely unmoved by their fate. And that is what eventually led to his imprisonment.

Once he was behind bars, he underwent — of his own volition — our neurofeedback training. By the time he had served his sentence, he was no longer a heartless thug. And he has not returned to his old pattern of behaviour to this day, three years later. He has taken up a mainstream profession and built a positive relationship with his sister. When he comes to visit us and speaks about his daily life and the people he deals with, his sweat glands show greater activity than before his treatment. This means that he is now more emotionally involved than before. He does not regret what he did before, but that was not to be expected, because his brain was not able to connect those actions with the idea of regret at the time he carried them out. On the other hand, he now no longer commits acts he might be expected to regret — and that is what ultimately counts.

## The trouble with reality

Despite the encouraging results we achieved with our neurofeedback treatment, there are, of course, limits to its realistic applicability when it comes to psychopaths. Although it can be seen as progress when they learn to increase the activity of their fear and empathy centres, for it to be useful it must occur successfully *all the time*. Those regions of their brains must be switched on beyond the confines of the lab, and not just whenever their owner is under scientific observation.

Most importantly, such people need to develop inhibitory activities to trigger when they are needed, such as when a repeat sex offender encounters a young girl he has sexual feeling for, or when a notorious thief sees an old lady collecting her pension at the bank. What is the use of a psychopath reacting with empathy and consideration under laboratory conditions if he falls back into his old behaviour pattern in the outside world?

Not to mention the fact that he — perhaps a highly talented trickster as many psychopaths are — may have been feigning empathy and scruples in the laboratory in order to be labelled 'cured' and released from prison.

In a nutshell: we have not yet been able to guarantee that every psychopath considered cured and released from prison will make that 'quantum leap' in real life, and I am in absolutely no doubt that giving such a guarantee is what is expected of us. Ultimately, the only way to give that guarantee would be to 'train' a treated psychopath gradually to re-enter daily life, and to maintain constant surveillance during that period. And even then there would be a residual risk of recidivism. In view of the potential danger to victims, that risk is understandably hard to accept.

A further limitation in the use of this technique to rehabilitate psychopaths is that it relies on an ability to persuade such people to undergo brain scans and practise controlling their brain activity. Patients with anxiety disorders and phobias are easily persuaded to do this because they want relief from their severely debilitating disorders. However, psychopaths are their precise opposites and so persuading them is not so simple. Usually, the only way is to offer them money. Lots of money! They have to be given a princely reward, and it has to come immediately, without delay, because psychopaths are unable to wait long. That makes the treatment expensive. On the other hand, keeping psychopaths in preventive custody for their entire lives is even more expensive. In the final reckoning, it is therefore probably cheaper to offer them a financial bait to participate in neurofeedback training.

Obviously, not everybody agrees with paying criminals to take part in their own therapy sessions. Yet it is unrealistic to expect someone whose brain lacks the physiological features necessary for ethical behaviour to decide to undergo treatment merely as a result of 'inner reflection' or a persuasive conversation. Anyone who does believe that that is possible is indulging in wishful thinking and ignoring the facts of brain physiology. However, those facts should not be taken as etched in stone, in a 'nothing-can-change-things' sense. Rather, it means accepting that psychopathy is mainly a condition of the brain, and the brain is plastic and malleable. Psychopathy is a fate, but not an inevitable one, as long as the brain is provided with the right learning conditions.

The same is true of severe degenerative conditions of the brain, which are commonly seen as resulting in the inevitable demise of a person's self and personality.

# 8

# The Trained Brain

*living better with Parkinson's and dementia*

Chimpanzees are lucky. They are spared the fate suffered by more than 50 million people worldwide — and probably twice as many as that by the year 2050: dementia. Apart from a few elderly zoo animals, you will hardly find an ape that can't remember where it stashed its oranges, or one that can no longer recognise its home, its relatives, or, eventually, its own self. Chimpanzee seniors are almost always still in full possession of their mental powers when they die, and their brains nearly always maintain the same volume as when they were in their prime and playing the mating game. They may become cantankerous, arthritic, hard of hearing, and half-blind, but their mental faculties are all present and correct. Enviable.

This begs the question of why chimpanzees don't get dementia. Zoologically speaking, they are very similar to us and their brains also work like ours in many ways: they think long and hard when they want to solve a problem; they are skilful at using tools; and they murder, lie, and cheat if it is to their personal advantage. Scientists at the Max Planck Institute for Evolutionary Anthropology, in Leipzig, have discovered that the genetic similarities between our two primate species are nowhere greater than inside our skulls: only 8 per cent of the genes in chimpanzees' brains are distinguishable from those in our own (whereas no fewer than 32 per cent of the genes in our testes behave differently, for example). Despite this, we are prone to Alzheimer's and other forms of dementia, and those apes are not. Why is that?

A trivial answer to this question was recently discovered by American anthropologists.[14] Simply: apes have a shorter life expectancy than we do — and this is sufficient to protect them from dementia.

The team of researchers led by Chet Sherwood at George Washington University used magnetic-resonance imaging to measure the brains of 87 human subjects between the ages of 22 and 88, and 99 chimpanzees between the ages of 10 and 51. They found that significant shrinkage in brain volume could be observed only in *Homo sapiens*, and the phenomenon was only present in subjects above the age of 50 — an age usually not even attained by the apes. The researchers interpreted this as clear proof that a stable brain structure and old age are basically incompatible. Or to put it another way: those who live on into old age — as is increasingly the case in industrialised nations — should not expect their brains to be able to perform in the same way they did when they were young. You can't have everything.

Evidently, the brain reaches a critical point at around 50 years of age: its cells no longer work as well, and this is accompanied by a build-up of damaging metabolites. As a result, neurons begin to die en masse, and, since replacement of those dead nerves is incomplete and does not reach all parts of the organ, the brain shrinks. Chimpanzees, however, are spared that fate. 'Their biological clock ticks faster, but the corresponding aging processes apparently do not affect their brains,' explains Sherwood. That's nice for the apes, and not so nice for us. But we should not despair, since there is also a good side to growing old, and our shrinking brain is usually sufficient to more or less get us through the day, and even still accomplish artistic and intellectual feats. Even when the degenerative process advances quickly, as in the disease generally known as Alzheimer's. This is down to the brain's mastery of the art of compensating (see illustration).

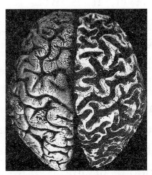

**Auguste Deter, the first dementia case described by Alois Alzheimer**

The hemisphere on the right is that of the dementia patient. The convolutions of the cerebral cortex are clearly shrunken, and the folds and fissures are clearly increased in size. As a comparison, the hemisphere on the left is from the healthy brain of a woman of the same age.

## Seahorses in trouble

Emaciated and wrinkled, full of holes like a Swiss cheese — this is essentially how we usually imagine the brain of an Alzheimer's patient to look, as if bits had been nibbled away from it. Or dried out like a sponge left out in the sun too long.

On the one hand, brain shrinkage is indeed one of the typical characteristics of dementia. On the other hand, it is a fate that none of us is spared, since several thousand of the cells of everyone's brain die every day once we pass the age of 20. Yet the consequences of this are minor at first. For one thing, the quality of a brain's functioning doesn't only depend on the number of cells it contains. For another thing, the reserves are huge, with more than 20 billion neurons and 100 trillion contacts (synapses). Thirdly, the brain's plasticity means it can compensate adequately for the loss of those cells.

However, it is well known that patients with Alzheimer's show some pretty severe symptoms. Initially, they experience problems with their short-term memory, then their learning and speaking abilities suffer, as do their fine motor skills, and eventually they are unable to recognise familiar people and objects, and their faculties for speech and self-reflection ebb away: patients lose their ability to communicate, and their sense of self. This crescendo of decline is, however, not only caused by an accelerated rate of cell death in the brain. It is also due to fact that the decline begins in a very particular part of the brain: the hippocampus.

The hippocampus bundles together individual

items of memory content to create a meaningful and understandable whole. For example, when you look at the face of your spouse, you see a forehead, two eyes, a nose, a mouth, a chin. But those sensory impressions are just the raw material, which then has to be connected to the memory content in your cerebrum to create a *Gesamtkunstwerk* if you are to perceive the face as that of your life partner. You first have to recognise that the moist ring of flesh encircling the teeth are 'lips', which means locating the concept in the memories stored in your cerebral cortex that corresponds to what you are seeing, and then you must also remember the special attributes that identify those lips as belonging to your spouse. So the hippocampus binds together the individual blossoms of our memory into a bouquet, which then causes the realisation in the cerebral cortex: 'Ah, that's my spouse.'

Patients with advanced Alzheimer's disease are no longer able to bind that bouquet together. They see the eyes, ears, nose, and chin, but can no longer associate them with a familiar person. In the most severe cases, they are no longer even able to access mental concepts for what they are seeing, so they can longer recognise lips as lips and a nose as a nose. They lose the associations and contexts they once took for granted.

It is not until cell death has set in in the hippocampus that this process also begins to occur in the cerebrum. The final result is the typically atrophied brain of a dementia patient that we know from many textbooks (see illustration on page 174). However, the reason why the cells of the cerebral cortex emulate the fate of their colleagues in the hippocampus has yet to be discovered.

If the flowers are not tied up in a bouquet, no one will want them. If the individual items of memory content are no longer bound together into a meaningful whole, they no longer have any meaning in themselves and they cannot be used — and anything in the body that is no longer used will atrophy, or even be deliberately destroyed to remove any 'dead weight'. This theory seems to be supported by a recent study carried out by researchers at Berlin's Charité hospital, in which Alzheimer's patients were found to have an antibody against certain receptors in the brain.[15] This antibody neutralises and inhibits associative information processing, which requires an intact connection between nerve cells to function.

The antibody discovery has led researchers to suspect that the condition known as Alzheimer's is an autoimmune disease, similar to arthritis, which causes strategic confusion among the body's immune defences, leading them to attack its own joints. This would mean immunosuppressive drugs might be effective in stemming the devastating immunological processes going on inside the skull. Or at least, they might be able to slow down the atrophying processes in the brain. However, it is equally possible that such medication has just as little effect as all the other drugs so far developed to treat Alzheimer's disease.

The pharmaceutical industry, with its innumerable army of researchers, sees dementia as a possible source of huge profits, not only because more and more people are suffering from the condition, but also because those with the condition have it *for many years*, making them into potential long-term pill-takers. This is why laboratories are working flat-out, but all their results so far have been disappointing. Meanwhile, pharmaceutical companies' profits continue to rise because their scientists and managers press tirelessly for early recognition of Alzheimer's disease, claiming that their drugs work better if taken early (for which there is not one shred of evidence).

Anyway, the search for an anti-dementia drug has not so far resulted in successful treatment. Presumably, this is because the ultimate aim of anti-dementia medication is to halt something that simply cannot be halted: the natural ageing process. There is no treatment against getting old, and that goes for the brain as much as the body. Nevertheless, the brain can be helped to compensate for the damage caused by ageing — with pretty good chances of success.

## Targeted training and stimulation

On the one hand, the fact that degeneration of the brain begins mainly in the hippocampus is devastating, because it hits right at the centre of our memory. On the other hand, this is also a region that forms the basis of the brain's plasticity, which primarily depends on the fact that individual items of memory content can be linked in a huge variety of different ways. The hippocampus is the mastermind behind that linking process — at least for spatial associations ('Where am I?') and explicitly conscious episodic memories ('My mother's name is …'). It goes without saying that the hippocampus needs to be extremely plastic in order to perform this function, which in turn means that even in the case of progressive degeneration, it retains many possibilities for maintaining its functionality.

But this is only the case if those resources are in regular use, and so the hippocampus requires training and stimulation. Most people are now aware of the fact that the progress of dementia can be slowed down effectively by remaining active for as long as possible. Buzzwords like 'brain jogging' and 'mental performance training' imply that the brain can be kept fit by constant use, just like any other organ of the body. However, the aspect that is often overlooked is the importance of *how* the brain is used. Busily solving crossword puzzles is of little help because, although that activity does involve extracting individual items of memory content from their mental drawers, it does little to help establish connections with other memory content, and thus with the hippocampus. Far better training effects are achieved through contextual learning, for example by integrating terms to be memorised into a story. And the training effects can be further heightened by creating a story that is patently absurd. So, to memorise the words 'plane' and 'giraffe', it is a good idea to visualise a giraffe sitting on a plane. This is an unusual association — and it's the kind of thing that trains the brain far more than solving crossword puzzles.

Indeed, anyone looking to train their brain should make a point of seeking out the unusual, in the sense of things they are not used to. When a computer specialist spends every day analysing bits and bytes to the point of exhaustion, or a chess grandmaster repeatedly plays out historic games in her mind, it will not do much to stave off dementia. It's like a veteran footballer who keeps on playing despite the twinges of pain in his knee — setting a course towards overstraining the ligament and making the condition worse, until eventually his knee is ruined. A better course of action would be to try out other forms of exercise, such as swimming or cycling, allowing him to go easy on his damaged meniscus while building up muscle and other tissue structures that will take the strain off his problematic knee. Not to mention the fact that it would also help keep him active and healthy. And this strategy of making activity as varied and flexible as possible, rather than repeating the same patterns over and over again, is also the one that achieves the best results for the brain in the battle against degeneration.

A study carried out at Boston University shows how this can be put into practice.[16] Thirteen Alzheimer's patients and 14 healthy older adults were asked to learn 40 children's songs that were previously unfamiliar to them. They were allowed to read the lyrics several times while receiving different kinds of support input: 20 of the song lyrics were accompanied by a corresponding sung recording, and 20 were accompanied by a spoken recording. Thus the visual stimulus of reading the text was associated with musical or spoken language stimuli. It may be assumed that some of the subjects would have had experience of the latter combination of stimuli, for example during a language learning course, but the subjects were unlikely to have experienced the first combination.

For the healthy participants, there appeared to be no performance difference in learning songs supported by spoken or sung recordings of the lyrics. The important thing for their brains was that they were able to associate the lyrics with some kind of stimulus, whether spoken or sung. By contrast, learning was much easier for the dementia patients when it was supported by musical stimuli. On the one hand, this corroborates clinical experience, which shows that the 'musical parts of the brain' are destroyed very late in the degenerative process caused by Alzheimer's disease; they survive intact for a relatively long time. On the other hand, it also shows that confronting dementia patients with brainteasers and learning tasks that force them to form new associations can be effective. Clearly, this stimulates the plastic reorganisation processes in the brain (see illustration on page 183).

Simply listening to music, however, offers about as much protection against dementia as watching television every day — that is, basically none. This is because it is a process that involves only consumption, and thus does not promote any kind of active learning or associative effort. Music must be combined with some kind of activity that demands associative effort of the brain. This may include not only learning song lyrics, but also performing motor tasks such as dancing or active music-making.

On average, active musicians develop Alzheimer's relatively late in life, and few conductors, whose profession means they are constantly moving to music, develop the condition at all. A number of studies have shown that a combination of active music-making and cognitive training (including brainteasers, memory games, and word puzzles) forms an effective strategy in the battle against various kinds of degenerative brain disease. Given the right instruction, patients can train independently and alone, but working in small groups has proven particularly effective. In either case, the exercises should be repeated as often as possible. This method can enable Alzheimer's patients to remain active in normal everyday life for around two years longer, which is a significant improvement considering that they can expect to live only another seven to ten years after their diagnosis. It is an improvement which benefits not only the patients themselves, but also their families and their care insurers.

However, it is of course also possible to achieve such an improvement without music, and even without tricky mental puzzles, as long as the brain is sufficiently stimulated. A team of researchers at the University Medical Centre Hamburg-Eppendorf motivated subjects between the ages of 50 and 67 to complete a three-month training course in how to juggle. The subjects' brains were measured before and after the training program. After the training phase, the scientists found an increase in the size of the area of the brain that specialises in motion and spatial awareness. They also discovered an increase in the size of the hippocampus. This was undoubtedly due not only to the fact that juggling places many different demands on the brain, from motor skills to coordination and spatial orientation, but also to the fact that their brains were engaging in a *new activity*, requiring them to create new neural pathways, which were completely or almost completely lacking before.

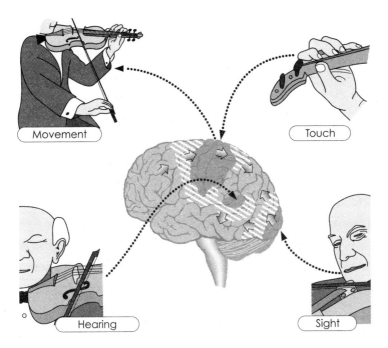

Movement

Touch

Hearing

Sight

### Music-making and the brain

Growth and increased connections in the brain (positive neuroplasticity) are especially promoted by active music-making because it causes complex acoustic (hearing), motor (playing, singing), visual (reading music, watching fellow musicians), and tactile (feeling the instrument) excitation patterns to converge simultaneously and in immediate succession in the temporal lobes and the shaded central areas of the brain, where they are connected with each other associatively.

Not included in the illustration but of equal importance are the simultaneous emotional activation and the logical connections created by reading or reproducing musical notation, as well as all types of memorisation and learning (conscious and unconscious) in the temporal lobes, hippocampus (conscious), and the basal ganglia (unconscious).

## Tuning the brain to the right wavelength

In Tübingen, we have carried out some successful work using a combination of cognitive training programs and electrical stimulation. This involves attaching contacts to patients' heads to pass certain electrical impulses through their brains. These impulses are not strong enough for the dementia patients to feel them, and thus they are not comparable with electric-shock treatments. However, they are strong enough to have a positive influence on learning.

To understand why, it is necessary to know that neurons communicate with each other in networks by building up an electric charge until they 'fire'. In an ideal learning process, many nerve cells will fire simultaneously, and if that happens often enough, the connections between those cells will become fixed — and the learned material is embedded in the memory. However, this is precisely the mechanism that breaks down in dementia patients: their neurons are no longer able to build up sufficient charge, and their firing is weak or non-existent. But it is possible to restart that process by providing the desired electrical charge from outside. This type of external stimulation is like giving a 'jump-start' to actions of the brain that it was designed to perform anyway: namely, building up electrical charges and transferring them between neurons.

This doesn't mean that such a jump-start is like an alarm signal, as if to say, 'a little electric shock to the head will increase alertness'. The effect of this electrical stimulation is rather the direct opposite. It often even produces the kind of brainwaves typical of deep sleep, as that phase of sleep is normally characterised by particularly intensive neural firing in order to fix permanently in the memory that which has been learned during waking. In small children, this phase can take up more than a quarter of a night's sleep, while for older people it is only very short, and patients with advanced dementia barely experience such a phase at all. But if the brains of patients with early-stage dementia are fed the necessary brainwaves by means of electrical stimulation, giving them back their deep sleep in a way, their memories can be given a jump-start — as long as they continue to train their brain as well. Electrostimulation cannot replace practice; it can only be a complement to it and help stabilise its successes. If nothing enters the brain for it to remember, it cannot practise memorising.

Training, learning, and electrical stimulation can all take place simultaneously if the stimulation takes place while the patient is awake. In the 1980s, when we first measured the slow brain potentials of our patients and research subjects while they completed various mental tasks, we noticed that performance was always improved whenever those potentials indicated a state of readiness in the brain. That realisation gave us the idea of trying to produce such a state deliberately, using weak electrical stimulation from two electrodes placed over areas of the brain responsible for certain functions. This was the birth of transcranial direct-current stimulation (tDCS), which is so popular today.

We carried out a large number of experiments using ourselves as guinea pigs, and some of them yielded results that were nothing less than philosophical in their dimension. For example, we administered the current to the left hemisphere, or the right, without the test subjects' knowledge. When they subsequently completed motor tests, they always used the hand on the opposite side to the side of their brain that had been stimulated, while their other hand remained virtually unused and the test subject showed no inclination to work with it. We had manipulated the subjects' will, which raises the question of how much 'free will' human beings can really be said to possess.

## Parkinson brains and champion compensators

Like dementia, Parkinson's disease does not strike patients out of the blue. Rather, it is a gradual process, which the brain can counteract to a certain extent. This may be seen by simply observing the progress of the disease. It is caused by the death of dopaminergic cells in the substantia nigra (a central area of the midbrain), causing ever lower levels of dopamine. This neurotransmitter is extremely important, because it is dopamine that makes us into creatures with drive and motivation. It is also necessary for motor control. This might lead us to expect that a lack of dopamine would quickly result in those infamous symptoms of Parkinson's disease, such as slowness of movement, rigidity, and shaking. Yet this is not the case.

Rather, although cell death in the substantia nigra often begins very early, sometimes as early as 35 years of age, without the patient noticing anything, symptoms don't usually appear until between the ages of 50 and 60, when 60 per cent, or sometimes even 80 per cent, of the dopaminergic cells have died. The brain manages to cushion the increasing lack of dopamine for many years, although the substance is as important as oil is for your car's engine. This is a masterly piece of compensation —

and if we could find a way to increase it, we would be able to slow the disease's progress even further than we now can.

Drugs used to treat Parkinson's, such a L-DOPA, have the opposite effect, because they shut down this compensatory process by providing the body artificially with dopamine and removing the need to compensate. This can even lead to the body eventually ceasing any residual dopamine production of its own, leaving it unable to contribute anything to the metabolism of this neurotransmitter. This throws the balance of transmitters in the brain out of kilter, and patients lose their last modicum of motivation and can often only sit in a wheelchair or lie in bed, apparently indifferent to their environment.

Of course, this kind of treatment can still be appropriate for older patients, as it offers them certain relief from other symptoms during the short time left to them. For younger patients, however, the approach must be a different one, since they can expect to live for many years with the disease — and so their treatment should make use of all the resources offered by their own bodies.

## When Parkinson's patients jump up from their wheelchairs

There are probably several aspects underlying the compensatory achievements of a brain damaged by Parkinson's disease. It is known that dopamine is produced not only in the substantia nigra, but also in other parts of the central nervous system. There is some evidence that those dopamine production areas kick in when the core dopaminergic region stops working. Another way of compensating involves restoring the balance of transmitters by reducing the production of dopamine antagonists such as acetylcholine. Furthermore, a team of German researchers has discovered that even before the appearance of their first symptoms, patients with Parkinson's disease activate significantly more of the motor areas of the brain to execute a given movement than do others. It is almost as if the brain decides to spread the burden of the task when it is weakened by a dopamine deficiency.

These are all plausible hypotheses, but ultimately, the mechanisms for these compensation strategies remain unclear. The important thing is that they work — and they work particularly well in the case of Parkinson's disease, although their effect is greater during the initial phase of the disease than in its more advanced stages.

This explains reports of Parkinson's patients suddenly leaping out of their wheelchairs and running away when immediate danger threatens, for example from a fire. Others show no emotion on their mask-like faces until their birthday party comes around, when suddenly they are all smiles, as if their illness had left them completely. For a long time, such incidents caused people to suspect that patients with Parkinson's might simply be faking it. What they actually are is proof of the brain's ability to compensate for losses — and proof of the fact that certain situations in particular encourage the development of those powers.

## A jump-start for the joker in reserve

Of course, we can't scare Parkinson's patients to death every time we want to get them on their feet, and it is not practical to celebrate their birthday every day in order to bring a smile to their faces. However, there are other situations and changes to their environment that can be used to encourage them to make use of the compensatory resources of their brains.

A line of tape stuck to the floor, leading from the patient's favourite spot to the toilet can often enable them to go to the loo alone, even after they are otherwise no longer able to do so. And, of course, a similar trick can be used to point the way to other destinations, such as the patient's bedroom or front door. The tape has the added advantage of stimulating patients to head off for these destinations independently.

Another example: patients with Parkinson's are often very wary of climbing stairs without a handrail, leading them to avoid stairs or to be reliant on help to negotiate them. But as soon as they see a handrail, they approach the staircase without trepidation. When their movements get stuck mid-flow, concentrating on the ascending line leading to the top of the staircase can help them get moving up the stairs again.

But why do strips of tape and staircase banisters get people with Parkinson's moving again? Is it that they give the patient a sense of security? This would somewhat contradict our example of patients running away from fire, in which the exact opposite — acute danger — causes them to move. An immediate hazard appears to mobilise huge powers for a moment of shock (secreting dopamine from their remaining cells), which however does not mean that patients can call on those powers at will in everyday life when they are not in shock. This leads us to assume that their reactions are based on something else entirely, which once again brings us back to the brain's constant desire to achieve an effect. The tape and the handrail — both bring patients closer to their goal by guiding them in the right direction. This indicates to the brain that the planned action will actually end in success, making it easier for the brain to make a connection between the journey and the destination, allowing it to access its reserves of dopamine to make the action possible.

Thus, tape on the floor does not get patients with Parkinson's disease moving because it indicates *where* the toilet is. Patients usually know that already. And despite the propensity of neurologists and psychiatrists to resort to the term 'Parkinson's dementia', patients do not necessarily always suffer from cognitive limitations. Rather, their problems lie in an inability to communicate with their environment at a normal speed, if at all. Tape on the floor gets them moving because it conveys the meaningfulness (the eventual effect) of that action and thereby causes the brain to activate its compensatory powers.

You could say that tape provides a jump-start for the brain to play its reserve joker. Just as a motorist whose car is running on the last drop of fuel is finally convinced to use the canister of petrol in his boot rather than saving it in the belief that it might be useful when times get tougher.

## Solidarity inside the brain

MRI neurofeedback can also help in these cases. Naturally, with Parkinson's patients, attention is focused on the areas of the brain that are important in motor function — not just those areas that are weakened, but also those that are able to compensate for the lost functions of the weakened areas. When patients manage to increase the flow of blood to those areas of their brain, they see this represented on a computer screen.

It is important that patients are given no instruction about *how* they are to increase the blood flow to those parts of their brain. They are free to choose what to think or feel in order to achieve this. It may be thinking of a dear friend, or imagining walking through a forest, sailing round the world, or watching a football final — it's entirely up to the patients themselves. Clinical experience has shown that those patients who are not really able to explain what was going through their mind when the anticipated symbol appeared on the screen are the best at directing blood to the desired areas of the brain. This is presumably because those areas are located deeper in the brain, beyond the reach of reflection and conscious memory.

A Dutch-British team of researchers achieved impressive results with MRI neurofeedback treatment for patients with Parkinson's disease.[17] The patients' motor function increased by 37 per cent after just two sessions, based on the Unified Parkinson's Disease Rating Scale, the international tool for measuring the stage of the disease. However, it remains to be seen whether this result can be transferred from the lab to everyday life. Nonetheless, the results are encouraging because they were achieved in only two sessions, which appears to indicate that the plastic powers of the brain can be activated relatively quickly. Furthermore, during this study, the blood supply increased not only to the target areas of the brain, but also to other regions that play a key role in initiating and controlling movement, including the subthalamus and the globus pallidus, which, in evolutionary terms, is part of the midbrain.

This once again shows that the individual areas of the brain can never be considered in isolation. When one weakens, others come to its assistance; and training one region necessarily means training others along with it. There are no lone wolves inside our skull, just team players all pulling together — unless they are being manipulated from outside. Unfortunately, this is precisely the strategy pursued by a psychiatric and neurological system dominated by the pharmaceutical industry.

Instead of pursuing that path, more funding must be approved and programs must be introduced to investigate scientifically the possibilities offered by neurofeedback for the containment of Parkinson's disease and dementia. Lobbying by pharmaceutical companies cannot be a basis for this, since, understandably, they have no interest in brain-training methods. So the patients themselves, as well as their families and advocates, must demand that such funding be provided. This has been far too rare an occurrence in the past. Patients and their families are rarely given information about neurofeedback because doctors generally do not believe that a physical disease can be controlled in this way.

# 9

# Fidgeting Is Not an Inescapable Fate

*treating ADHD without drugs*

It starts right after birth. The baby barely sleeps, and so neither do his parents, of course. He cries, can't stop moving, and appears to be suffering in some way. But the infant cannot be diagnosed with any recognisable condition. Even the infamous 'three months' colic' cannot be blamed as the cause. Although the child's sleep might improve somewhat, an unbroken night's sleep remains an impossible dream right up until he reaches school age.

The child begins to get into trouble at kindergarten: he is impulsive and is always getting mixed up in minor disasters; he damages toys, usually unintentionally, is clumsy, and disturbs other children's play with his volatile nature. He is easily distracted and is liable to change what or where he is playing abruptly. The preschool teachers are unable to control him, and can only sigh in despair. Other children's parents complain about his behaviour.

Once in primary school, he does not fare any better. The child is 'annoying' and defiant, challenging both teachers and fellow pupils. Although he is just as able to solve problems as any other member of his class, and shows the same level of intelligence, he lacks concentration and perseverance. Everyone can see this child is different. First signs of aggression appear without warning and are often totally unbridled. The resulting sanctions from teachers and fellow pupils lead to self-doubt, low self-esteem, and increasing social isolation. The child grows lonely, and the foundations of a later 'career' as a psychopath, addict, or criminal have been laid.

ADHD, attention-deficit hyperactivity disorder, is far more than just trivial naughtiness. But it also need not be a catastrophe. Even Albert Einstein, the 'archetypal' absent-minded professor, is thought to have had ADHD. The same is true of Wolfgang Amadeus Mozart, Thomas Edison, Thomas Mann, John Lennon, John F. Kennedy, and even the 'great soul' of the struggle for Indian independence, Mahatma Gandhi. At least, that's the prevailing opinion circulated widely in the media and on the internet, often accompanied by quotes from experts in the field. This alone should be enough to reassure parents of children with ADHD. After all, winning a Nobel Prize, becoming president of the United States, or being recognised as an icon of modern music are not bad prospects for a child's future. Still, many parents are heartbroken by such a diagnosis and often begin to panic.

Then they ask their doctor about the drug

methylphenidate, better known by the trade name Ritalin. After the failure of countless other types of therapy, they are simply at their wits' end, and Ritalin is their only hope for their child to finally start to sit still and concentrate in class. They hope it will mean their child is no longer bullied for being a crazy weirdo and that he will get on better with his teachers and his classmates. They also dream of him getting better grades. And not least of all, they look forward to what they have missed since his birth — a normal, relaxed family life. After all, children with ADHD are a huge challenge for many parents.

For pharmaceutical companies, such a scenario is a boon. Nothing is better for their sales figures than a mental disorder that mainly affects children, has parents worried sick, and destroys both family and school life, but which can also be controlled instantly with a pill. The amount of methylphenidate prescribed in Germany, the home of Ritalin, has sky-rocketed in the past 20 years, increasing by 184 times, to the equivalent of just under two tonnes of the drug a year. Global sales of Ritalin earned the Nuremberg-based pharmaceutical company Novartis $464 million in 2010. Just four years earlier, that figure stood at $330 million.

In view of such developments, we must wonder whether the number of children with ADHD really has shot up in recent years, or whether it is simply the number of diagnoses and prescriptions of Ritalin that have risen so sharply. It is in the interests of both the pharmaceutical industry and therapists to declare a disorder to be a mass phenomenon, since this increases the income of both.

Be that as it may, we must also bear in mind that there *are* children who have ADHD. But it remains questionable whether they should be treated with an amphetamine-like drug that is banned in sport, is chemically related to speed and cocaine, can cause addiction, impairs the growth and movement of children, and is regulated by narcotics legislation when given to adults. Especially since ADHD can also be controlled without the aid of pharmaceuticals — using the brain's plasticity.

## It all starts with a screaming baby

The first indications that a child has ADHD often appear very early. Children normally develop a stable sleep–wake rhythm by the age of three to six months, and no longer need to be carried around constantly and coaxed into sleeping with all manner of tricks. But for parents of babies with ADHD, there is no relief and the ordeal goes on. The baby simply refuses to go to sleep, and, when it does, it wakes up repeatedly, and usually lets its parents know by screaming as loud as it can. There are cases of ADHD developing later in childhood as a result of parenting or other environmental influences, but the disorder usually manifests itself in infancy.

In an analysis of 22 studies, including almost 17,000 children, researchers concluded that babies with so-called regulatory disorders — excessive crying, sleep disturbance, reluctance to feed, late success in toilet training — are 40 times more likely to develop an ADHD-type behavioural disorder later in life. Although such statistics say little about cause and effect, they can be seen as a clear indication that children with ADHD have a huge problem with control: they are impulsive and easily bored; they ignore important but common signals while reacting disproportionately to unusual stimuli; and they are not easily pacified after such a reaction.

The fact that this behaviour manifests itself so early in childhood — before the end of infancy — could be taken as an indication that the causes of ADHD are mainly genetic in nature, leading to the assumption that nothing can be done about it and parents should be thankful that they can at least keep it in check with Ritalin. But that reasoning is not necessarily sound.

The brain's plasticity means that even mental characteristics with a strong genetic component can be shaped. Furthermore, other studies have shown that babies who cry a lot are shaken more often and more roughly by their fathers and mothers, and so it could be the case for many such children that the problems they encounter later in life with concentration and learning are exacerbated by neurological damage sustained as a result of such parental violence. And shaking is probably only the tip of an iceberg of acts of parental frustration that undoubtedly have an effect on the child's development.

It remains unclear whether the causes of ADHD are genetic or social, or a mixture of both. The data available at the present time merely confirm that the brains of children with ADHD develop regulatory disorders very early in life. Their lack of impulse control and their desire for powerful stimuli make them reminiscent of a type you have already met in this book: psychopaths. You shall soon see that this resemblance is not coincidental.

## Aggression in school

Children with ADHD usually make their first contact with other children in kindergarten, and this is normally also where they first encounter authority from adults other than their parents. Conflict is inevitable. Those around them will refuse to tolerate a child with ADHD's demands for attention or lack of obedience. The parents are then blamed for not bringing the child up properly, or they are already advised to put their child on medication — that is, to tranquilise their child with Ritalin.

All this does not exactly serve to calm the situation. Indeed, it rather serves to confirm the children's own feeling that there is 'something wrong' with them. Such a self-perception breeds frustration in those children, and leads to a lack of self-confidence, which is often expressed as aggression towards others. Aggression and hostile behaviour are frequently quoted as typical symptoms of ADHD, but they are often simply a result of the frustration experienced by those with the disorder at the lack of acceptance they are met with from the people around them.

Once they enter school, the situation of children with ADHD escalates further. As they grow bigger and stronger, teachers and other staff can no longer easily physically control them. Moreover, school children are expected to be able to concentrate and focus on their work, and this is one of the things children with ADHD struggle with most. Studies have shown that children with ADHD achieve lower grades and are more likely to leave school without qualifications than their peers.

As teenagers, these children show a preference for high-risk or even criminal behaviour, such as driving without a licence or fare-dodging on public transport, and have a tendency to steal spontaneously. Furthermore, they start smoking and regularly consuming alcohol and other drugs earlier than other teenagers, and become sexually active relatively young: we know that boys with ADHD are implicated in teenage pregnancies far more often than the average (and 60 to 80 per cent of ADHD patients are male). Their sexual precocity, increased tendency towards addiction, and affinity towards criminal activity are more ways in which children with ADHD are similar to psychopaths.

## The call for the cure-all pill

Schools and their staff are usually unable to cope. The call for Ritalin becomes more insistent. Many parents, their resistance ground down by the constant fruitless arguments with their child and with those who complain about their child, eventually give way and allow their child to be put on drugs. Prescription rates are soaring around the world. In Germany, around 200,000 children of school age take Ritalin; in the United States, that number has reached more than six million. The peak age for prescriptions is between nine and 11 years; however, in the US, this medication is given to six-year-olds, and sometimes to children as young as four. Furthermore, a significant increase in dosages has been observed recently. Daily doses of 60 mg and more are now no longer a rarity, although that is equivalent to or even more than the recommended highest dose, and the drug is normally prescribed in doses of 10–20 mg per day.

There can be no doubt *that* Ritalin works — that is, it tranquilises children and increases their concentration. However, *how* it does this seems surprising at first, since, as a relative of amphetamines, it actually works as strong stimulant. One possible explanation is based on the biological principle of homeostasis, the process by which every organism strives to achieve balance of powers within itself. This means that the already agitated body and mind of an ADHD patient does not respond to Ritalin by pushing the body further — possibly to the point of complete exhaustion — but instead slams on the emergency brake and sends out signals to calm the body and mind. It is also known that coffee can be an effective agent for 'grounding' hyperactive children with ADHD, in a way we might more readily expect of valerian or Valium. Pharmaceutical tranquilisers do not help and can even worsen the symptoms of ADHD, by increasing a patient's urge to reach a higher level of agitation.

However, experts have now moved on from the theory that Ritalin works in this quasi-homeopathic way, fighting like with like. Today, doctors believe Ritalin works by inhibiting the uptake of the neurotransmitters dopamine and noradrenaline in the synaptic cleft, which is the point of contact between nerve cells. This allows the transmitter substances to continue to do their job of linking neurones together rather than 'getting lost' inside them. This effect strongly activates the part of the brain that languishes in standby mode in the brain of ADHD patients: the prefrontal cortex. This is the area responsible for controlling actions that may otherwise be inappropriate in specific social situations, and for emotional regulation. In ADHD patients, activity in this part of the brain is relatively subdued, and it receives too little blood flow. But Ritalin, with its effect on the balance of neurotransmitters, jolts it into action. The result is that the ADHD patients are better able to concentrate and control their actions, limiting them to those which are socially acceptable to people around them.

So far, so good. However, Ritalin has many side effects, some of which can be quite serious. For example, it can inhibit growth: children who take Ritalin tend to be smaller than their peers. Furthermore, like almost all psychotropic substances, Ritalin does not give the brain support only where it is needed, but floods the brain with neurotransmitter substances, which affect parts of the organ where this is not desirable. Typical symptoms resulting from this include insomnia, headaches, nervousness, anxiety, depressed mood, tics, dizziness, and teeth grinding. Generally, the longer the drug is taken, the stronger the side effects become. These also include the risk of addiction, which is often played down by pharmaceutical companies. It should not be dismissed out of hand, however, since ADHD patients who receive Ritalin are precisely those whose condition means they have a heightened risk of addictive behaviour.

## From fidgety child to psychopath

This is all compounded by the fact that the positive effects of Ritalin are rather limited. Ritalin has little effect on ADHD patients once they pass puberty. And it does not influence the number of children with ADHD who later go on to become psychopaths, drug addicts, or career criminals — ironically, this is the effect that would be most needed. Studies have found that one in every five alcoholics showed clear signs of ADHD in retrospect; among those addicted to other substances (such as heroin and cocaine), the rate was even higher, reaching up to 50 per cent. We do not know whether these people are induced to try drugs by their impulsive nature or as an attempt to self-medicate their disorder — but there can be little doubt about the connection between ADHD and drug consumption. Likewise, there can be little doubt about the connection between ADHD and criminal behaviour.

An examination of the inmates of one prison in Scotland showed that 23 per cent had clear signs of ADHD in their childhood and one in five of them displayed symptoms even as adults. Forensic scientists noticed, however, that the numbers can vary widely if inmates are sorted according to the type of crime they committed. The percentage is relatively low (13 per cent) among those imprisoned for traffic offences, such as hit-and-run drivers and those caught in charge of a vehicle without a licence; among those incarcerated for sex crimes and robbery, the rate is extremely high (31 and 35 per cent respectively). These figures mean those convicted of serious crimes especially will often turn out to have a history of ADHD and, in some cases, still struggle with the disorder as adults.

So there are more than enough reasons to seek a long-term method for treating ADHD. Ritalin turned out not to be the answer (which is perhaps unsurprising in view of the addiction problems often suffered by such patients — someone who has learned from childhood on that drugs can be a source of help will always carry that behaviour pattern in the back of their mind as an option). Moreover, the pronounced similarities between ADHD and psychopathy indicate that drug-based treatments are likely to be of limited effectiveness: as yet, there is no medication to reliably treat psychopaths, and the same must be assumed for ADHD patients, who display not only similar behaviours but also similar anomalies in the working of their brains.

Of course, parents need not worry that their fidgety, unfocused child will one day grow up to be a coldblooded psychopath. ADHD can also be associated with extreme creativity, highly original ways of thinking, and unusual personalities — just think of Einstein, Kennedy, Lennon, and the rest. There is no proof that feeding a child Ritalin will stunt a potential career of that calibre by smothering any spontaneity, creativity, and emotional responsivity — but it is conceivable.

And there is no reason to despair of any treatment working at all. Options exist that do not tranquilise kids with drugs, but which rely on teaching them how to stimulate their own brains in such a way that it develops the desired ability to concentrate and react in a socially appropriate way. We have already shown that the brains of adult psychopaths can certainly be reshaped. Psychopaths often like to bend the truth, mislead, and generally toy with other people — including their therapists — so it may not be an easy task, but it can work, in principle. And we can safely say that it is all the more effective as a treatment for children with ADHD, since their young brains are particularly plastic and almost beg to be shaped. They also have a natural urge to play and cooperate, which any therapy should make use of before these children grow out of it.

## Self-control through neurofeedback

Neurofeedback therapy offers a promising approach here. It aims to teach ADHD patients to control the activity in their own frontal lobes in order to improve their attention and concentration. In this procedure, a computer analyses the patient's brainwaves, separating them according to their frequencies and visualising them in graphic form on a screen for the patient to see (see illustration on page 206).

Children have no inherent interest in abstract representations of brainwaves, so the graphics should be playful in nature, such as a flying plane whose altitude they can alter by influencing their brain activity. Or a ball they have to shoot into a goal in the same way. They would score a goal every time they managed to increase their slow brain potentials. This is effective in treating attention deficit because it pre-activates those neurons in the frontal area of the brain that play a significant role in concentration, self-control, and control of socially appropriate reactions. That means those cells are prepared electrically for the task they will soon have to complete.

Another possible treatment involves using neurofeedback to help patients learn to increase the amplitude of a particular type of brainwave called the sensorimotor rhythm (SMR). The SMR is an oscillation that protects the brain system from interference factors. It appears in the region of the brain's motor areas and has a regular frequency of 8–15 hertz and is a signal that the excitation circuits connecting the cerebrum and the diencephalon (interbrain) are ready for action. Any movement — even imagined or observed movement — reduces this wave range; learning to increase it, however, can reduce a patient's level of inattentiveness, hyperactivity, and impulsivity.

We do not currently know exactly which ADHD patients benefit more from increasing their slow brain potentials and which benefit more from increasing their SMR. Since neither of these training programs results in negative side effects, however, further research can proceed. Anyway, in both cases, the aim is to teach young patients to control and redirect their own brain activity to achieve a certain goal on the screen and inside their brains. It's like a computer game, only rather than pushing buttons, players have to control what happens on the screen directly, using their neuronal activity. However, the young patients are left to decide for themselves how to influence the events on the screen — that is, what thoughts they conjure up in order to manoeuvre the ball into the goal.

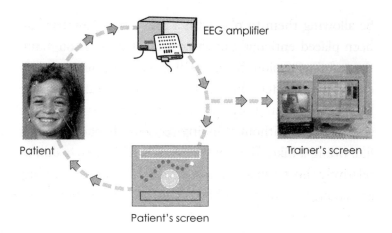

Patient's screen

### Neurofeedback self-control training for ADHD

The young patient (left) watches a white dot 'float' across the screen (bottom). The patients' job is to manoeuvre the dot into the white goal at the top in just a few seconds using only their brainwaves. If successful, they are rewarded with a smiley face and a point. The points can be traded in for toys after the training session. The EEG amplifier (top) transfers the brainwaves from the patients' frontal cortex to the patient's screen and the trainer's computer (right). Patients are left to decide for themselves how to manoeuvre the dot into the goal. They may think of something specific, call up certain emotions, or simply do nothing in particular. The only condition is that they may not move their eyes.

On the left, the young patient can be seen wearing the EEG electrodes — small metal patches attached above his frontal cortex and his eye sockets. Learning to increase the frontal, so-called 'slow brain potentials' in particular results in improved self-control and concentration on learning tasks.

In the next phase, the screen remains black, and the children no longer see a game — no flying plane and no ball and goal. They are asked nonetheless to think of exactly the same pathways in their brain as they did to win points in the game. If they manage that, flying blind so to speak, they receive a real reward. For small children, this might be allowing them to play with a toy from a box that has been placed enticingly in their field of view throughout the training session. Such positive reinforcement teaches children to switch their brains to concentration mode by themselves, and, by the end of the process, they are able to do so without thinking consciously about it, just like riding a bike. The time needed to achieve that skill is relatively short. On average, as few as 13 hours of training are enough to teach children how to switch their own brains to concentration mode.

A case study: eight-year-old Maximilian was highly intelligent (IQ 130), but his scores in attention tests were significantly below average, and he was having great trouble at school, in particular with arithmetic and any activities that required extended periods of concentration. Maximilian's father, a medical professional, was aware of the risks associated with Ritalin and therefore wanted his son first to try the less dangerous method of neurofeedback training. Maximilian now learned how to increase the slow brain potentials in his frontal cortex by watching a space rocket on a screen and trying to make it fly in a particular direction using those brain potentials. When he steered the rocket to the right-hand side of the screen, a smiley face lit up and a pleasant bell-like signal sounded. If he managed to do that ten times, an exclamation mark on the screen told him he would be allowed to choose a toy after the training session. Before each training session, he was told that he should not move during the process, and he soon realised that physical movement interfered with the rocket's flight. The only other instruction he was given was that he should think of 'something'.

Children quickly learn to control their brains; they are not usually able to express how they do it, but, after a few hours of practice, they can feel that their concentration has improved and their inattentiveness has lessened. Maximilian underwent attention tests after every five sessions. Even after just ten sessions, his scores were completely normal, and reports from his class teacher were also much more positive. Of course, he still needed to learn to implement this procedure when it was actually necessary — i.e. whenever he was required to learn or solve cognitive problems. This is also achieved using positive reinforcement: children are rewarded whenever they manage to learn something or solve a problem by switching to their personal concentration mode. And, of course, there is also always the incidental reward of achieving more success in such activities than before. The procedure becomes automatic and permanently changes the brain so that patients are always able to implement the thought strategies they learned during neurofeedback training.

## Like Ritalin, but with a future

Research has shown that the therapeutic effects of neurofeedback are comparable with those of Ritalin: the same increase in concentration and abilities, the same improvement in social behaviour, the same calming effect. But unlike the drug, neurofeedback has no side effects. Furthermore, neurofeedback is more likely to result in permanent improvement, since learned behaviour patterns can survive puberty — and even when they don't, they can always be re-learned.

Neurofeedback using an EEG is a *more-indirect* way to achieve the aim of activating the frontal areas of the brain, in which the increased appearance of slow brain potentials is taken as an indicator that those brain regions are doing more work. The successes gained with this treatment vouch for the usefulness of this approach, but more precision would be gained if the mobilisation of the frontal cortex could be observed in the metabolic activity of the brain. Fortunately, near-infrared spectroscopy makes this possible.

This method is also especially suitable for those whose disorder includes a high degree of hyperactivity. The infrared-LED cap is no more cumbersome than a bathing cap, and children can even walk around while wearing it and learning to produce the blood-flow patterns in the brain that are associated with inhibiting movement. (It should be noted that many children with ADHD only become so fidgety because they are unable to cope with their condition and express their frustration in impulsive or even aggressive movements. In such patients, hyperactivity is not a primary symptom but a side effect of their disorder — and neurofeedback with an EEG can help here once again.)

In Tübingen, we have already achieved considerable success in treating ADHD with neurofeedback. It offers a real alternative to treating these disorders with Ritalin and similar drugs, which has been the dominant approach until now. The most important advantage of neurofeedback treatment is that it does not rely on manipulating the brain with pharmaceuticals, but rather helps the brain to help itself in order to free it from the functional disorder it is struggling with. This not only reduces the risk of side effects, but also increases the probability that the newly gained functionality of the brain will remain stable. Most of the children we treated for ADHD still had relatively good concentration abilities two years later, and if the therapeutic effect should begin to weaken, it can easily be reinforced with more neurofeedback sessions.

By contrast, Ritalin has only been proven to be effective in the short term; no study has shown any long-term effects that reach beyond puberty. Much points towards the possibility that ADHD is far from gone once the medication is stopped, but rather results in the appearance of new problems, such as withdrawal symptoms, anxiety, and depression. Not to mention the fact that Ritalin and related substances can impair the maturation process of children and adolescents. The effects of those drugs are limited to that which is pharmaceutically possible: a person learns nothing new by taking medication.

# 10

# Genius for All

*how we can improve our perception*

Kim Peek was an unusual man, although at first it did not seem likely that he would ever come to anything. At birth, his skull was one-third larger than that of other babies, making it too heavy for his neck muscles to support and causing it to flop forward all the time. Although he was not diagnosed with the notorious condition hydrocephalus, Kim's doctors warned his parents not to hold out much hope for their child's future development. In fact, they considered him to be a hopeless case, arguing that his disability meant that the best place for him would be a care home.

And indeed, Kim did grow up to be a rather awkward and clumsy child, learning to walk and talk much later than normal and never really learning to do up his own shoelaces and shirt buttons. The boy with the oversized head preferred to spend every day sorting scraps of paper, reacting hysterically when he was disturbed. His age peers considered him a misfit, although that did not seem to concern him much. He lived in a world of his own, first learning to read and then remembering everything he read. His feats of memory were truly amazing!

At the age of four, he was already able to recite the contents of eight volumes of an encyclopaedia almost verbatim. Kim's father realised that his son would require a very different kind of educational support than normal. He solved crossword puzzles with his son every day and provided him with piles of books and newspapers to read. After Kim was expelled from school for his 'uncontrollable' behaviour, he was homeschooled by tutors. The strategy worked, and Kim completed the high-school curriculum by the age of 14. Despite this, his performance in IQ tests was varied at best. Depending on the type of test and the skills it focused on, Kim could score as high as 184 — Einstein — or as low as 72 — cognitive disability.

At a church service when he was 12 years old, Kim recited 40 verses of the Bible, despite the fact that he had never read them and had heard them only once. He began to catch the attention of the public, and talk of the extraordinary young man soon reached Hollywood. He became the real-life inspiration for the title character in the film *Rain Man*, portrayed by Dustin Hoffman as a strange but likeable autistic man with extreme savant talents, who won the hearts of many a movie-goer. While researching his role, the Hollywood star spent a day with this genius of mental recall and advised Kim's father, Fran, to show his son's talents off to a wider audience. This would help the young man come out of his shell and bring some money in, which was sorely needed since, as a single father and Kim's full-time carer, Fran Peek had no time to hold down a paid job.

Fran was hesitant at first. He believed his son would not feel comfortable outside of his familiar environment. But eventually he began accepting invitations for Kim to attend interviews and thrill audiences at schools and universities with his memory talents. This turned out to have a beneficial side effect: Kim's social skills improved greatly. The Peeks became a competent father–child team: the child was the performer, the father was the manager, and both benefitted from the arrangement. Despite efforts in the media to paint their relationship as exploitative on the part of Fran, there is no doubt that Kim flourished in the face of the new demands his fame placed on him. And anyway, he was not autistic, despite Dustin Hoffman's portrayal of him as such. He was perhaps 'weird' and unable to care for himself without the help of his father, his way of communicating was stilted and rather unorthodox, and he sometimes needed time to himself in order to concentrate on things no other person would be even remotely interested in. But otherwise, he sought out the company of his fellow human beings, hoping to earn their respect and understanding, just as anybody would.

Also, Kim loved to amaze audiences with his spectacular feats of memory. He was able to say spontaneously what day of the week any given date fell or will fall on, and he could give the zip code for any address in the United States. He could recite from memory all the leaders of Germany since Bismarck, and identify all the musical instruments playing a piece of orchestral music after hearing just seconds of a recording. Kim had the contents of a total of 12,000 books stored in his memory. He did not understand much of that information, but he could rattle it off on command. It took him around seven seconds to memorise completely two pages of any book.

Of course, neurologists were curious to know what lay hidden beneath Kim's oversized skull. Using functional magnetic-resonance imaging, they discovered not only a particularly voluminous cerebrum, but also found that it was almost completely unconnected to the lower levels of his brain. His cerebellum appeared to be unusually small, which might explain Kim's motor problems. The most striking feature of Kim's brain, however, was that his corpus callosum, which normally connects the two hemispheres of the brain, was almost entirely missing — and when the two halves of the brain are unable to influence each other and keep each other in check, it may be that information flows unhindered into the consciousness.

All this can only partially explain Kim's abilities. But it does feed our prejudice that savant syndrome, sometimes called 'islands of genius', can only occur when something goes wrong in the brain. This prejudice also allows us to comfort ourselves with the fact that we may not have a brilliant talent, but at least we are not sick in the head.

However, the fact is that only 50 per cent of savants are diagnosed with autism, and among the remaining 50 per cent, there are many people who may seem strange to us because of their very specific talent, but otherwise think in more or less the same way as anybody else. This raises the question of whether any one of us could be at least a little 'savant'. If we are honest, most of us would often be glad of an ability to remember facts as easily as Kim did and to complete complicated calculations by switching on our brains rather than a pocket calculator. And which of us has never secretly longed for the ability to impress others with some extraordinary talent? It may seem unlikely, but in fact it is not impossible for mere mortals like you or me to develop such an 'island of genius' — at least, the possibility is provided by our brain.

**Drawing by a three-year-old autistic savant**

This excellent artistic drawing of a horse displays an extraordinarily high level of ability for a three-year-old, in one isolated area. Highly developed artistic, memory, or calculation skills are particularly common.

## Rapid access to the preconscious

Our team in Tübingen studied the phenomenon of savants and examined autistic people with 'island of genius' talents. We discovered that their brains are activated much more quickly than normal when they perceive something. The main areas of the brain that were mobilised were those responsible for 'pre-attentive processing', the early, preconscious processing of signals that occurs within the first 100 milliseconds of perception. These areas are distributed throughout the brain, depending on which stimulus pathways are taken. In perceiving acoustic signals, they will be in the auditory cortex on the upper sides of the temporal lobes; for visual signals, they will be in the visual cortex at the back of the cerebrum. Combined stimuli — which are the norm — can activate several areas at once.

What is of more significance, however, is that in the brains of savants, it is not conscious, filtered perception that dominates, but rather preconscious, unfiltered perception. Such people have easier access to that which is currently being filed away in the memory but has not yet become conscious knowledge. However, the next process, which takes place 200–300 milliseconds later, is underdeveloped in autistic savants.

An example from everyday life might help to demonstrate what this means in real terms. Imagine you are driving through town in a car. At a crossroads, the lights suddenly turn red. First, that signal passes through the preconscious level, within 100 milliseconds, which is enough time for you to instinctively assess that stimulus as a sign of potential danger — you immediately slam on the brakes. This results in your car skidding to a halt, and, only then, while the tyres are still screeching, do you consciously register the red light. This is because it takes more time and a wider spread of excitation in the brain for a signal to reach your consciousness, requiring at least 200 and perhaps even as long as 300 milliseconds. You draw a deep breath and sit back in the car seat. That was a lucky escape!

But the lucky thing is that our brain reacts to stimuli and assesses their significance *before* we are conscious of them. This was useful in the evolutionary fight for survival, and it remains useful to this day. The only difference being that we used to have to beware of wild bears, and now our challenge is to survive the traffic.

The brains of autistic savants have particularly good access to the preconscious. However, this doesn't mean that they react extremely quickly. On the contrary, they are anything but skilled survivors, and would probably slam on the brakes much more slowly than the rest of us in the red-traffic-light scenario. Yet depending on their particular talent, they might be able to say instantaneously how many red traffic lights there are at the intersection, or how many cars or pedestrians are currently there. Or like 'Rain Man', they might know immediately how many matches fell out of the box when the car executed that emergency stop. They may have the ability to count things faster than we can slam on the brakes at a red light. But they lack the filter that allows them to sift the important information from a flood of stimuli — in this case, the information that they should step on the brakes because the light has turned red. Instead, they register unimportant things — or rather: things that seem unimportant *to us*. And they do this with a degree of perfection that we, in turn, find hard to believe.

One way to put it is that the perception window of savants is more open than ours. On the other hand, they have far fewer windows than we do. Kim Peek was able to rattle off the names of all the World Series baseball players of the past hundred years, but had trouble recognising the faces of the people around him. He also never really learned to clean his teeth, and generally paid little attention to his own health. He died of a heart attack in 2009, at the age of 58. This may have been due to the fact that he was a capacious eater and never physically exercised.

Another savant, the Russian journalist Solomon Shereshevsky was far less fortunate. Due to a brain defect, he was unable to forget anything, no matter how meaningless the memory might be. He was tormented by this, was unable to continue working in his profession, and had to earn a living working as a circus mnemonist. He searched desperately for techniques that would enable him to forget. For example, he tried writing down lists of numbers and burning them — but the numbers just appeared to him on the charred remains of the paper. Solomon eventually descended into depression and delusion. He died in 1958, at the age of 70.

The connection between delusional mental illness and unfiltered perception is borne out by the fact that schizophrenia patients often have photographic memories. It is difficult to define cause and effect in this phenomenon — but there is clearly good reason for a person not to keep their perception window too far open under normal circumstances. This would lead to perception overload.

It is no coincidence that savants are often found to be on the autism spectrum. Autism often brings with it a large number of cognitive, social, and linguistic deficits, leading to the development of only a very limited number of wide-open perception windows, leaving people with autism susceptible to being overwhelmed with too much information.

## A little bit savant

However, this is no reason for us not to try to open up our personal 'savant window'. In our research at Tübingen, we examined both autistic and non-autistic people with a savant ability. We found that their brains functioned in a similar way in terms of priority of perceptions, having similarly rapid access to the preconscious. The EEG readings and MRI results left us in no doubt. However, unlike our autistic test subjects, those without autism were not swamped by this flood of data, but were able to filter out the most important information in a given situation. This demonstrates that 'islands of genius' do not necessarily go hand in hand with a severe brain disorder. It also means that it must be possible for those with healthy brains to develop savant abilities, at least to a certain extent.

One way to achieve this makes use of the technique

of magnetic brain stimulation. A study carried out by the Spanish-American brain scientist Alvaro Pascual-Leone shows how this can work. His test subjects first listened to a story being read aloud, which they were expected to retell later in the experiment. While they were listening to the story, some subjects had the frontal and temporal lobes of their cerebrum stimulated using transcranial magnetic stimulation (TMS), which exposes the brain to a pulsating magnetic field. The other subjects had typical TMS coils attached to their heads, but these were not active, functioning only as a placebo. The result was that those whose brains had been electrically stimulated were able to recall 15 per cent more details when they retold the story.

The Australian researcher Robyn Young used repetitive TMS on her test subjects. This involves 'firing off' a rapid succession of magnetic impulses at a rate of 10–20 hertz. She concentrated on the areas of the brain that are responsible for preconscious perception. Five of her 17 test subjects did indeed develop a savant-like ability as a result: they were suddenly able to remember any calendar dates, or they developed surprising new artistic abilities.

It must be remembered that the transcranial magnetic stimulation had no effect on 12 of the 17 test subjects in the Australian study. Another issue is that such procedures smack somewhat of manipulation: someone switches on a machine and alters the person attached to it. This not only raises moral issues but also begs the question of how stable the effects it achieves can be — that is, how long the effects will last after the magnetic coils are removed from the subject's head.

## More attention through meditation

Other ways of opening our perception window are offered by meditation techniques such as those used in many Asian schools of philosophy and religion. Their basic principle of achieving a state of being in the moment with no other purpose, and of absolute awareness of that moment, already includes many aspects of pre-attentive perception as described above.

The Japanese Zen master Takuan Sōhō once said: 'When facing a single tree, if you look at a single one of its red leaves, you will not see all the others. When the eye is not set on one leaf, and you face the tree with nothing at all in mind, any number of leaves are visible to the eye without limit. But if a single leaf holds the eye, it will be as if the remaining leaves were not there.' This unfixed, pure way of viewing a tree as a whole is very reminiscent of the wide-open perception window of the savant.

Nonetheless, I used to be highly sceptical of yoga, meditation, and other supposedly consciousness-expanding exercises. This was partly because my team and I had been unable to see any special brain activity represented in the EEG readings of meditating test subjects. Their EEGs looked like those of someone in normal, worldly sleep rather than indicating some kind of transcendental consciousness. However, it appears that this was due to the fact that our test subjects were beginners rather than experts in meditation techniques. If you attach experienced yogis or Zen masters to an EEG machine, however, the results look very different. Still not transcendental, but different to those of a sleeping person — and not only in terms of their brainwaves.

When, in 1992, the Canadian psychologist Jane Raymond showed her test subjects a rapid succession of letters on a screen, she found out that they could only reproduce the sequence perfectly if there was an interval of at least 500 milliseconds between each letter and the next. When two letters appeared in quicker succession than that, the second one was almost always completely missed: as far as the subject's conscious mind was concerned, the second letter never actually appeared. After Raymond's results became known, the idea that human beings are unable to perceive visual stimuli appearing in very rapid succession came to be seen as the basic constant of human perception. But, in 2007, it was shown that this can vary.

The Dutch neuroscientist Heleen Slagter repeated Jane Raymond's test with 17 subjects.[18] The difference was that the subjects were tested twice; once at the beginning of the study, and then again after an intensive (eight to ten hours a day!) three-month training course in meditation. In the second test, every subject was able to perceive the second letter. The 23 people in the control group, who had received just one hour's introduction to meditation techniques and had practised occasionally by themselves at home, were not able to improve their so-called 'attentional blink' deficit. The second letter still slipped through their perceptive net.

To investigate why the meditating subjects were able to increase their perceptive awareness to such a large extent, Slagter hooked them up to an EEG. The brain potentials she measured showed a less pronounced response to seeing the first letter, which also continued for less time than before. This means that, before seeing the second letter, the subjects' perception level had almost returned to its initial state. This might indicate that the subjects had regained sufficient neuronal capacity to perceive the second letter consciously. Or to put it another way, they were better able to apportion their awareness resources because they reacted less intensely to the stimuli they received — and that, in turn, might be due to the fact that they really were now seeing the letters individually, without their brains trying to arrange them into some structure, context, or hierarchy of meaning.

What we can learn from this is that meditation exercises can help to open the brain's momentary perception window. Furthermore, compared to technology-based procedures, meditation has the added advantage of enabling people to push *themselves* towards developing savant abilities. It is not others activating savant capabilities when they see fit, but the person him- or herself, acting of their own volition.

Meditation techniques are not everybody's cup of tea, however. Various studies have shown that such techniques as yoga or Buddhist meditation only work when the practitioner has the requisite motivation and beliefs. Without those, practitioners will not be sufficiently disciplined meditators, and the results will be the same as those registered for Slagter's control group of occasional, self-guided meditators, whose perception windows did not open even a tiny bit wider. Nevertheless, other procedures remain available to those hoping to activate their savant potential despite being sceptical about meditation. One of those techniques is neurofeedback.

## More attention through neurofeedback

As with other areas in which neurofeedback is applied, the procedure here is based on the principle of users watching the workings of their own brains in real time. Again, the representation of this can be very simple: for example, a red dot on the screen, which the participant is asked to attempt to turn green. Participants will then try anything they can think of: clicking their fingers, thinking of the last time they had sex, humming a tune, or remembering the time they failed that crucial school exam. What they don't know is that the colour will only change when they manage to activate that part of their brain that's responsible for perception within the preconscious timeframe of up to 50 milliseconds. Eventually they will manage to change the dot's colour by chance, and then they must try to repeat that success by 'tuning' their brains to the same 'wavelength' as before. As a rule, people require just two, one-hour training sessions to learn to activate the relevant parts of their brain at will.

## When the unconscious controls itself

Using savant brain training, my colleague Sunjung Kim proved that healthy people can learn to control their preconscious and unconscious minds (see illustration on page 225). First, she presented her test subjects with images of various faces on a screen, with either happy or sad expressions. However, the faces appeared for a period of 16–100 milliseconds, while conscious perception usually takes 100–300 milliseconds. So, although the test subjects were in fact constantly looking at happy or sad faces, they claimed to have seen nothing.

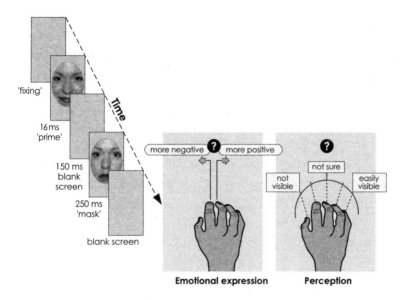

**Sunjung Kim's experiment on conscious and preconscious perception**

A blank 'fixing' screen was shown first for 6 seconds, then a positive or negative facial expression (called the 'prime') appeared for such a short time that it could not consciously be perceived (in this case, for 16 milliseconds). The screen then went blank again for 150 milliseconds, before a neutral face (the so-called 'mask') appeared for 250 milliseconds and could easily be recognised. Then came another short pause (the sequence is shown on the left of the illustration). After that pause, test subjects had to use their right hand to rate the facial expression of the 'mask' as positive or negative, and then rate the visibility of the first 'prime' (shown on the right of the illustration).

Kim then presented her subjects with an image of a face for a longer duration, so that they could consciously perceive it. This face was neither happy nor sad, but completely neutral and indifferent in its expression. Basically, a poker face. The subjects were asked to evaluate its expression — and they always picked the same emotion as that which they had just 'seen' before, even though it was below the threshold of their consciousness. When they had unconsciously looked at sad photos, they judged the neutral portrait to be sad; if they had unconsciously seen happy photos, they believed the portrait to be displaying a happy expression. Their unconscious experience had coloured their conscious perception.

It is not difficult to see why politics and industry are interested in subliminal methods of manipulating our moods and opinions. Luckily, it was made illegal in most countries many decades ago — as the result of an experiment carried out in movie theatres in 1957 by James Vicary, in which cinema-goers were induced to buy popcorn and Coca-Cola by hidden advertising messages. However, this study later turned out to have been a hoax. It was not until this century that studies proved that subliminal advertising can work if it matches with the recipients' desires at the time.

In the next phase of her experiment, Kim trained her — completely non-autistic, healthy — subjects to increase the blood flow to those areas of the brain responsible for perception within the preconscious time window, using the savant brain-training method described earlier (see illustration on page 26) while in a magnetic-resonance scanner. Her results were astonishing: all subjects who had learned to strengthen those neural systems (and *only* those!) were able to consciously perceive the faces and emotional expressions that had been invisible to them before the training, even if they were shown them for only 15–30 milliseconds. Such a reduction in perception time had never before been observed in healthy people! Kim then trained the same subjects to reduce the blood flow in the relevant parts of their brains, and their perception times returned to their original level.

What would Freud have said if he had been told that it does not take hundreds of hours of psychoanalysis sessions to become conscious of our unconscious, but just two, one-hour neurofeedback sessions? This is all the training Kim's test subjects required. Furthermore, her experiment showed that we are able to learn to control our subconscious subconsciously. None of the subjects were able to say how they increased the blood flow to their brain's consciousness system, although the savant effect was clear — 'islands of genius' for all is not just a pipedream.

## An option for autism therapy

Neurofeedback is not only able to tease savant abilities out of healthy people; it can work in the other direction as a therapy for autism — regardless of whether the person has savant characteristics or not. For it to be used, the patient's brain activity must first be carefully analysed using visualisation technology, since autism is a many-faceted condition. Patients who display particularly severe compulsive behaviour and fixation on certain actions often show increased beta-activity in the brain, while others who are highly impulsive and hyperactive are more likely to show pronounced theta and delta activity. Different wave patterns require different neurofeedback approaches: patients with compulsive behaviour need to reduce their beta waves; those with impulsive behaviour must reduce their theta and delta activity; and so on.

Initial studies have now been completed into the effectiveness of neurofeedback in treating autism. An improvement of up to 40 per cent was observed in symptomatic social avoidance behaviour, whereby the success was always greatest when patients were required to reduce the activity in a certain region of their brain or in the connections between specific brain regions.

Persuading people with autism to undergo treatment in the first place remains a major challenge. They often avoid social contact and new experiences, preferring to remain in their familiar environment. A plus point for neurofeedback therapy, on the other hand, is that it mainly requires the autistic patient to interact alone with a computer, without the need for any intensive communication with other people, which is a circumstance likely to suit autistic patients. As with the treatment of hyperactive children, neurofeedback training can be designed as a game, in which, for example, patients try to manoeuvre a car or a spaceship through a labyrinth using the activity of their brains. In this way, neurofeedback training is experienced not as treatment, but as play.

## No brilliance at the push of a button

Despite the great therapeutic successes achieved using neurofeedback, I would like to stress that it cannot perform miracles. It will not turn anyone into a genius overnight in an area they have no talent or training in. Someone who has never been able to play the piano is not suddenly going to become a virtuoso at the keyboard just by activating the relevant brain regions. Brilliance at the push of a button is not going to happen. Research into the nature of talent has shown that great musicians such as Mozart or Beethoven, and also John Lennon and Michael Jackson, all accumulated around 10,000 hours of practice by the time they were 20 years old. This shows that genius only comes to fruition as a result of dedication and diligence, and must first manifest itself in early childhood.

For this reason, it makes sense for us to awaken our savant potential in an area in which we already have some talent. A person who is already a relatively accomplished guitarist can train her preconscious mind so that she can think less about planning her playing and can allow the music to simply flow from her fingertips, so to speak. This can be especially good for music genres such as jazz or blues, in which improvisation plays an important part. That is one side of the coin.

The other side is that opening the savant window can lead to a musician disappearing down the rabbit hole of his or her own attentiveness and blocking out the rest of the world. This can hinder communication with an audience or with fellow musicians, creating a temporary state of 'musical autism', which is a drawback in particular at live concerts with a band or orchestra.

How far are people willing to go to develop their potential? Even the Ancient Greeks complained of people's *pleonexia* — their self-indulgence and greed. We always want more than is good for us or our environment. More talent, more money, more power. More love, more sex, more entertainment. Why are we unable to just sit back and be satisfied with what we have? Why must we always want more and more, even when it defies all reason? Once again, the answer to that question is to be found in our brain and its matchless capacity for plasticity.

# 11

# Nirvana

*is there life beyond greed, addiction,
and want, want, want?*

Bunga bunga! It's a phrase that did the rounds in the media a couple of years ago and has now passed into general usage. No one is really sure where it came from or what it originally meant. It seems to have first appeared around the turn of the last century, in the context of British colonialism, perhaps as a reference to the immoral goings-on in the harems of African and Middle Eastern rulers. It is clearly meant to sound as if it comes from a 'primitive' language of the jungle or the bush. This was a way of reinforcing differentness: wild savages with their unbridled sexuality, contrasted with the decent, upstanding colonial overlords, whose moral superiority gave them the absolute right to rule over such barbarians. But few people these days are aware of this historical background when they hear the phrase 'bunga bunga'.

Today, 'bunga bunga' conjures up images in our minds

of old men cavorting with much younger women. Men like the octogenarian Silvio Berlusconi, who did not stop appearing in public with his young female playmates even after a court sentenced him to seven years in prison for paying for sex with an underage girl. Or like the nonagenarian Hugh Hefner, who still likes to surround himself with Playboy bunnies and who, on New Year's Eve 2012, married a woman young enough to be his granddaughter.

In this way, 'bunga bunga' describes an inability to stop doing something even when it's no longer in a person's best interests, or, worse, when it is associated with disadvantage. Like a smoker who can't kick the tobacco habit, these men are unable keep their hands off young women, even in extreme old age, and even if their behaviour makes them look ridiculous in most people's eyes.

I do not intend to stoke any kind of moral outrage by mentioning this, since no one has the right to stick a sell-by date on older men's sexual activity, or to demand that, if they really must have sex, they'd better do it with women of their own age. But even a free spirit like the American pop artist Andy Warhol once claimed that true freedom only comes once you're through with sex. So it seems legitimate to ask why the Berlusconis and Hefners of this world don't finally go into sexual retirement and enjoy the wisdom and liberty of old age. And, more generally, why so many people find it impossible to stop pursuing their desires and addictions although they know from experience that they will not provide ultimate satisfaction, but rather a short-term kick until the next urge comes along.

This behaviour pattern is not confined to old men continually chasing after young women and girls. When financial speculators unhinge the entire global economy with their rampant risky dealing, it is once again a case of 'not being able to get enough', which can refer equally to the actual money and the adrenaline rush they are chasing. The unbridled desire for power displayed by dictators and absolute rulers, which costs the lives of millions of people every year, also fits into this category, of course, as do the countless numbers of people who are trapped in addiction.

Strongly physically addictive drugs like heroin play only a minor role here, as psychological addictions are far more widespread. Innumerable people are hooked on gambling, interactive computer gaming, and the internet. Others' eating habits are out of control and they are unable to stop themselves, even when their obesity leads to health problems and social isolation. Others become stalkers, violating the privacy of the object of their obsession.

There appear to be no limits on the variety of addiction — we now even have many cases of 'tanorexia', the obsessive need to lie in the sun or on a sunbed until the skin is more reminiscent of tanned leather than human tissue. Other addictions even seem to have become generally acceptable: when the actresses in ads for online shoe retailers shriek ecstatically over their newly delivered shoes, the implication is that shopping addictions have become socially acceptable. At least here in Europe.

Lists of possible addictions go into the dozens or even hundreds, depending on who draws them up. The German Professional Association for Addiction puts the number of people in Germany who are addicted to drugs and alcohol at 3.2 million; 3.8 million are thought to be addicted to nicotine; and then there are the non-substance-based addictions, to shopping or gambling, for instance. The German government's drugs commissioner speaks of 16 million smokers, 1.3 million alcoholics, 600,000 regular cannabis users, and more than 500,000 gambling addicts, as well as an estimated 8 per cent of users who are addicted to the internet. However, none of those figures are particularly reliable. This is firstly because experts still argue over what exactly should be labelled an addiction. Secondly, very few people are addicted to only one thing, so there is much overlap. Thirdly, the statistics gathered about psychiatric and psychological disorders in general are rather vague. 'In Germany, pigs are counted three times a year, and the fruit harvest is enumerated several times annually,' laments the Düsseldorf-based sociologist Karl-Heinz Reuband, 'but the number of people admitted every year to hospitals or psychiatric institutions with particular diagnoses is not recorded in any statistics.'

Nonetheless, we can be certain that addiction is a mass phenomenon. Addiction are such a mass phenomenon that it begs the question of whether many of them can still be called pathological, or whether they have become a part of 'just the usual craziness'. When we take a closer look at our brains, it seems we must conclude that this is most likely the case.

## The liking hardly changes — but the wanting increases

Arthur Schopenhauer believed the will to be 'the eternal and indestructible in man'. Although he believed there were differences in the concerns of younger and older people ('The character of the first half of life is an unsatisfied longing for happiness; that of the second is dread of misfortune'), and he recognised the possibility of affirming or denying the will, he believed that, in itself, the will was both infinite and everlasting: 'All philosophers have made the mistake of placing that which is metaphysical, indestructible, and eternal in the intellect. It lies exclusively in the will.'

Schopenhauer's great philosophical insight also holds true from the point of view of brain research. The will — in the sense of desirous wanting or wishing — is the anticipation, the expectation of the effects that result from our behaviour, and so it is indeed indestructible, and it also forms the basis of almost every learning process. However, in recent decades a distinction has established itself that offers a better explanation for why ageing playboys continue to seek the company of younger women in their old age, or why elder statesmen insisting on chain-smoking despite their doctors' orders. We now distinguish between *liking* and *wanting* — and while the things we like change only negligibly, our will to get them increases continually (see illustration on page 237).

An example from ordinary life might help explain this distinction better. There are many different things we might like. These can include sunsets, going to the cinema, beautiful faces, or sausages with sauerkraut. But we don't feel the need to do everything in our power to have or experience these things as often as possible. We have no difficulty doing without them for a couple of days or even weeks.

However, in principle at least, any of these things can potentially become the object of a stronger desire, which could eventually become so powerful that we feel we can no longer do without them. Sunsets and sausages with sauerkraut can become an addiction just as much as cigarettes and alcohol. Due to the psychotropic substances they contain, which — unlike sunsets — *directly* affect the systems in the brain that control wanting, cigarettes and alcohol do pose a greater risk; but in principle we can develop an irresistible desire that even goes as far as a dependency to anything we like.

## Anyone can become addicted to almost anything

Let's assume you're a fan of beautiful sunsets. You share this love with other people, and nobody thinks anything of it when you pull your car over of an evening to admire the picturesque scene from the roadside parking area, as the red orb in the sky slowly descends below the horizon. For most people, this would be enough to satisfy their desire to experience the spectacular wonders of nature for quite some time. But for you, watching the sunset at the side of the road is more than just a pleasant interlude; it triggers intense feelings of euphoria.

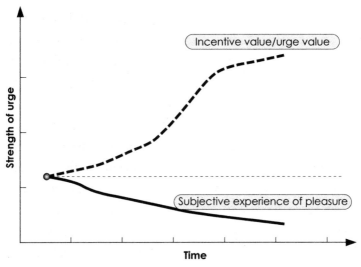

**The development of 'wanting' and 'liking'**

This graph shows the development of 'wanting' (wishing for, top curve) and 'liking' (bottom curve) over time after repeated positive stimuli (e.g. from consuming drugs) or behaviour leading to positive feelings of satisfaction (e.g. having sex).

The attractiveness (or the 'incentive value') of positive stimuli increases exponentially with each repeated experience, while the subjective experience of satisfaction slowly declines: the latter phenomenon is often described as habituation. The rapid increase in the desire to experience a given stimulus is associated with the increased release of the neurotransmitter dopamine in the brain.

And so, the next evening, you drive to another roadside layby to repeat the experience one more time. It works, and once again you experience pleasure and sensual satisfaction. And so it goes on, until, eventually, laybys are no longer just laybys, but hold the promise of intense feelings of delight and euphoria. You now feel a thrill of exhilaration whenever you approach one, just like the thrill a shopaholic feels when she enters a boutique, or a porn addict feels when he switches on his computer. The actual event, the sunset, no longer takes centre stage, and is replaced by the layby that holds the promise of a beautiful sunset.

Boutiques, computers, and roadside laybys are nothing special in our modern world. Often, we don't even notice them, and, even if we do, they have little significance for us. We make use of them whenever we need to, but, beyond their pragmatic purpose, we are indifferent to them. However, if our brain connects them with a powerfully pleasurable stimulus, they increase dramatically in significance for us. This is what psychologists call salience: the stimulus is taken out of context, losing its neutrality and taking on enormous meaning. It's like a spot of red colour that jumps out at us from a sea of blue. The layby is no longer just a random place by the roadside, the boutique is no longer just a dress shop, and the PC is no longer just a work tool; they now harbour the promise of euphoria — and it is to that promise that we are liable to become addicted.

Such an addiction adversely affects people's behaviour in social and work situations by causing such an obsession that their perception is dominated by it. They spend the whole day waiting for the evening; they can think of nothing else, and refuse all invitations that do not include a sure promise of a sunset — until their whole lives revolve around sunsets. If no sunset happens to be available at a given moment, they enthuse with other sunset addicts online or engage in replacement activities, such as collecting photos or paintings of sunsets. If they can't get their regular fix, due to days of extended cloud coverage, they become increasingly nervous and restless and unable to concentrate. Eventually, they might start to film sunsets in order to guarantee a constant supply and to enable them to enjoy the spectacle even during the hours of daylight. Old friends and acquaintances find this obsession difficult to understand and begin to turn their backs on the sunset addict in irritation.

Of course, the addicts themselves become increasingly numbed to the effects of sunsets as time goes on, and the scenes have to be ever-more intense and impressive, which prompts the addict to head off to ever-more remote corners of the Earth. And the pinnacle is an evening flight heading west, which offers a permanent sunset to admire. All that travelling eventually leads to financial ruin, social isolation (except from fellow addicts), and unemployment, after getting caught by the boss ogling sunsets on the computer at work ...

Of course, the chances of becoming addicted to sunsets is relatively small, not least because we get to see them too rarely. By contrast, the danger of becoming dependent on online chat, video games, or pornography is much greater because such addictions satisfy vital and fundamental human needs, such as the need for communication, adventure, excitement, or sex. Theoretically, however, almost any positive, and even some negative experiences and stimuli can become the object of an addiction — in the case of anorexia, this can even be the person's own body, or rather the negation of it — because these things anyway become less and less significant as the addiction develops. After a few fixes, even heroin can no longer provide the initial euphoric effect on its own, but still the addict's body and brain crave it. At that point, the anticipation of the pleasurable experience, the expectation of the moment of fulfilment, has become more significant than the act of satisfaction itself — and this is precisely the vehicle on which it is possible to slide into addiction to practically anything.

Thus, addictions do not come about, as believed by many people — including many doctors, psychologists, and addiction therapists — as a result of the negative stimulus of withdrawal (for example: as a result of the shakes and nausea we experience after a night of heavy drinking, as the alcohol level in our blood sinks, leading us to reach for the bottle to alleviate those symptoms). Such a mechanism does play a supporting role, but of far greater importance is the fact that the incentive value of the situations associated with the object of addiction increases, even when the object itself is no longer so attractive.

This is the reason why 80 per cent of addicts relapse within a year of withdrawal. An alcoholic may emerge clean from the clinic, but the next time he walks past a bar, he will still be drawn as powerfully to enter it as he was before, which is why the danger is so great that he will dismiss all good intentions and go in and get drunk. A smoker may be determined to cut cigarettes out of her life on the stroke of midnight on New Year's Eve, but as soon as she finds herself in day-to-day situations in which she would routinely light up, such as with that first cup of coffee and the newspaper in the morning, or in front of the TV with a glass of wine in the evening, she will suddenly find herself facing a huge problem: her brain telling her in no uncertain terms that the pleasure of that morning coffee or that evening glass of wine is only complete with a cigarette. *These* are the situations in which people relapse — and no one, even with the best will in the world, is immune to them: from a homeless heroin addict to a rich and successful football manager such as Germany's Uli Hoeness, who, during his tax-evasion trial, admitted to rampant, unbridled gambling on the financial markets. He ignored the risk of losing his entire fortune and, as a person in the public eye, the danger of becoming a national laughing stock in exactly the same way as heroin addicts ignore the physical decline of their bodies. The power of positive association is too great to allow such fears to gain the upper hand.

In the case of Uli Hoeness, the accusation was often made that a man of his experience should never have reached the point where he was permanently glued like a teenager to his smartphone or laptop, speculating on the financial markets. But there is no reason to hope that such urges, and thus the risk of addiction that goes along with them, will become less pronounced as we get older. On the contrary, although satisfying such wants happens less, the wanting itself actually increases. This explains why there are so many men whose impotence means they can no longer have sex at all, but who still gawk at any woman in a short skirt. There are enough examples from history to prove that lust for power increases continually over the course of the years and takes on ever more brutal forms, and this doesn't apply only to dictators such as Hitler or Stalin.

Also, the idea that addictions appear mainly in younger people is wrong. In Germany, 1.5 million people over the age of 60 are addicted to sleeping pills and tranquilisers, and this is not an addiction they have had since their younger years, as is often the case with smoking, but rather one they have developed in old age. It happens according to the pattern described above: in the beginning, there is the wonderful experience of finally being able to get a good night's sleep, with the aid of benzodiazepines; later, the brain can no longer imagine sleeping without them, although the body's physical habituation processes mean they no longer have very much pharmaceutical effect.

## Better dead than without dopamine

The great importance of joyous anticipation and association in the development of addiction can also be described in neurobiological terms, in particular in terms of a certain functional system: the mesolimbic dopamine pathway. It is made up of cells with long axons originating on the border between the midbrain and the interbrain and stretching all the way to the frontal regions. When this pathway is stimulated, it can trigger an urge to act that overrides any feelings of caution, even to the point of self-destruction.

In one experiment, hungry rats were placed in a T-shaped labyrinth. Food was available in the branches of the T, and in the longer section there was a lever that they had been taught would electrically stimulate their dopamine system when they activated it. It should not be forgotten that these animals were hungry, and the path to available food was short and easy.

However, once the rats had understood which reward was to be found where, they went exclusively for the electric stimulation. They pressed the lever up to 5,000 times a day, until they were too exhausted to press it any more. It was as if they would rather die than live without their dopamine fix. They even abandoned their young and their sexual partners, and were prepared to jump electrified dividers to get their fix. They also continued to do so even when researchers denied them the reward of the electric impulse for a time: the mere anticipation and association of the reward stimulus with the corresponding branch of the T-maze was enough to make them try to get there.

Observations of the dopamine systems of people who are addicted to nicotine show that these are considerably more active *before* they get their fix of the drug than during intake. And their dopamine system becomes especially active if the drug is not administered at the expected time. What this means is that dopamine levels are among their highest *before* the final object of wanting is attained, and reach an even higher level *after* the want fails to be satisfied, but there is no surge in dopamine release while the drug is being consumed. This explains why a smoker who is still smoking a cigarette will fidget like a nervous squirrel if he realises the packet is empty. It has absolutely nothing to do with withdrawal — after all, our smoker still has a lit cigarette between his lips. Rather, this reaction comes because the smoker's neurobiology is now absolutely tuned to wanting.

## What pubs and churches have in common

Against this background, we must consider whether addiction should really be seen as a medical issue at all, since recognised dependencies such as alcoholism or gambling addiction are controlled by the same circuits in the brain as romantic love and strict religious belief. These control circuits mainly function like this: there is something that we like (it is not possible without this — no one can become an alcoholic if they don't like alcoholic drinks or the effect they have!), and there are situations that increase the feeling of wanting it. For alcoholics, this may be the hubbub of a crowded bar or the fun atmosphere at a party; for the pious believer, it might be the sight of a church or actual contact with a sin. Ultimately, however, the two behaviour patterns boil down to the same thing: a learned motivation in anticipation of something that is experienced as pleasurable.

The fact that dependencies on drugs like heroin, nicotine, and alcohol can cause sickness and death is in no doubt. What is doubtful, however, is whether an irresistible desire for something should be generally branded as pathological. By that logic, married couples who have nothing more to say to each other after 40 years together but do not break up, because neither one can imagine life without the other, should be labelled as sick. Eating breakfast or watching TV alone is unthinkable for them. Just as the smoker's brain cannot bear to do without that postprandial or post-coital cigarette. Some widowed partners actually suffer from extreme withdrawal symptoms, and many soon follow their spouses to the grave. Their behaviour is described as unbreakable loyalty, while smokers are branded addicts and forced into glass cages at airports, the likes of which are usually only seen imprisoning the defendants at show trials in totalitarian countries. This does not seem very consistent. For this reason, the World Health Organisation (WHO) speaks of a dependence requiring treatment only if it significantly impairs the person's daily life and leads to health problems. But it also expressly states that these circumstances can also be caused by such things as love, jealousy, and monetary greed. This insight has not yet made its way into the canon of psychiatric practice or addiction therapy.

We must accept the fact that every brain has an enormous potential for addiction. This is the dark side of the phenomena we have been examining so closely: the brain's constant quest for effects (especially its pursuit of rewards) on the one hand, and its great plasticity on the other. The quest for effects (anticipation of rewards) sparks interest in the brain and triggers action; the brain's plasticity means it can focus on almost anything as the object of that quest. This object may be dangerous or even threatening to the brain owner's very existence — as shown by the experiment with the rats who preferred to get a dopamine fix than to fill their empty stomachs. But brains don't necessarily think about the repercussions of their actions or the best interests of their owners — what they care about is achieving the effects they crave, the immediate satisfaction of a need or desire.

## Escaping addiction

A harmful addiction must, of course, be treated if the person in question is unable to learn self-control. A behaviour-therapy-based option is offered by aversion therapy, based on the principle of conditioning. Consumption of the addictive substance is deliberately associated with extremely negative stimuli. This might be a harmless but painful electric shock; for those trying to quit smoking, it might also be a deliberate overdose caused by smoking a large number of cigarettes in quick succession to the point of nausea — with the process being repeated several times a day. This 'rapid smoking' is, however, only effective if it takes place not only in a scientifically controlled environment such as a psychotherapist's practice, but also in the natural environment in which the nicotine addict would normally smoke. As I already described, addiction therapies must focus on the situation in which the object of the addiction is consumed.

This is why sending teenage drug addicts to the countryside for treatment is unlikely to lead to long-term success. They may indeed stay clean for a while during their stay on a farm, but as soon as they leave that neutral location and return to their usual environment and social circles, every street, bar, park, club, or public toilet will remind them of their past life. Their old friends and acquaintances are still using drugs as they did before, increasing both availability and peer pressure to particularly high levels, and so the young addict in treatment is advised to stay well away from them. But this will probably lead to isolation and confrontation with the fact that they have become part of the 'boring mainstream', which they always professed to hate. Constant sobriety does not necessarily make the world a more interesting place; on the contrary, it is hard work resisting the urge to get high. Especially if you know that all you need to do is go round the corner to the nearest club and snort a bit of coke to escape the misery. Very powerful positive incentives for self-control are necessary to prevent a relapse in such a situation. So much so that it seems almost miraculous that some people manage to remain abstinent.

Withdrawal is only the first step. Smokers also face the question of how to deal with situations in which they would normally light up a cigarette. What are they supposed to do while they are at their desk, waiting for the computer to boot? Or when they're stuck in traffic, or waiting for the bus? How can they escape the peer pressure of their still-smoking friends without putting their friendships in jeopardy?

This is why addiction therapy has the greatest chance of success if all the 'dangerous' situations — that is, those that are associated with the object of the addiction — are removed, or if the length of time between their occurrence is made considerably longer. Moving to a neutral area, not associated with the addiction, can help and is worthwhile, but not always realistic. New activities, habits, and routines to fill the vacuum left by the object of the addiction are enormously important.

Yet such radical life changes are not easy — and they are rare. In addition, they are often incompatible with the life habits of the addict. A computer specialist who is addicted to the internet can't simply avoid computers, since that would rob him of his livelihood. In such a case, the aim must be to restore the computer to the position of simple working tool. The only allowable association must be: computer = work, and not: computer = pleasure. This necessarily means that such a person must find another way to relieve stress, occupy his free time, and provide pleasure. A difficult task, but not impossible. At least the brain possesses the necessary plasticity to achieve it.

## Can wanting be extinguished?

With this in mind, would it not be better to curb the wanting before it reaches the excesses of addiction? Buddhism is well-known for advocating this, its ultimate aim being the extinction of the will in the form of nirvana — the state in which there is no desire, no attachment, and no self-centred activity.

In western philosophy, the main proponent of this position was Arthur Schopenhauer. His philosophy was that life is eternal suffering and that this is the fault of the will, or of 'wanting', affecting the deep unconscious. This is because it is wanting that sparks any action in life, any interest we have, our wish to eat and drink, to desire each other and reproduce, to dominate and demean each other. The will is always desirous of fulfilment, and as long as we do not achieve that fulfilment, we will feel pain, which Schopenhauer also sees as including the feeling of being driven, which becomes stronger the more the will is forced to struggle against resistance. However, if the object of fulfilment of the will is achieved, it will not be long before we are overcome with boredom, until the wanting re-emerges, and the whole process begins again from the start. Thus, pain, lack of freedom, and boredom are the substance of existence, which is why Schopenhauer claims, 'Every life story is a story of suffering'. This may sound pessimistic, but I believe Schopenhauer would have felt vindicated by the findings of modern brain science.

The same excitation curve of the mesolimbic dopamine pathway as described for drug addicts is seen in healthy individuals in a less pronounced form. It also rises rapidly *before* an aim is achieved, only to level off sharply once that objective is attained. We know that Parkinson patients who are no longer receiving dopamine agonists suffer similar symptoms to an alcoholic who is no longer getting anything to drink, and the basic principle is true of healthy people, too. When healthy individuals want something, they often become rather nervous and preoccupied, focusing their attention on how to attain the object of their desire. Once that goal has been achieved, they experience a brief moment of fulfilment, only to become despondent soon after. Self-doubt and depression set in, the previously driven and driving desire gives way to sadness and resignation. This cascade of suffering, fed by the activities of the dopaminergic system, sounds decidedly Schopenhauerian.

Incidentally, Schopenhauer would presumably have

had no problem with his philosophy finding support in the results of modern brain research; after all, it was he who said, 'Just as for the motion of a ball upon impact, so also ultimately for the brain's thought, a physical explanation must be in itself possible'.

However, Schopenhauer was not a total pessimist; he did recognise opportunities to escape the eternal tribulation caused by wanting. He completely ruled out suicide, since that is a decision driven by the will. In fact, when people are treated for depression, either with drugs or psychotherapy, and initial signs of success begin to appear, that is often when they kill themselves. The initial treatment succeeds in awakening their previously dulled will, giving them the energy they previously lacked to go through with suicide. According to Schopenhauer, the preferred aim must be to extinguish the will, like snuffing the flame of a candle.

Schopenhauer saw one way to achieve this as being through art, by viewing the world from the standpoint of the artist, which should be disinterested and therefore liberated from will. A flautist himself, the philosopher considered music in particular to be an instrument of such deliverance, because in music we experience not things, but the will itself directly, allowing us to transcend the level of the objectifying want altogether. This theory is underpinned by the fact that research into addiction has never been able to identify an addiction to music as such. For sure, many musicians are workaholics or drug addicts, but this is more likely a consequence of the high-pressure industry they work in, as well as the narcissistic and psychopathic tendencies of many star performers. All those people who continually expose themselves to sounds through their personal stereo, smartphone, radio, or television could be described noise addicts, since they cannot stand silence around them. However, music itself appears to have no addictive properties.

According to Schopenhauer, other opportunities for deliverance from wanting are offered by self-denial and by compassion. Self-denial, as a way of breaking the power of wanting; compassion, to overcome one's own, subjective suffering through empathy with the suffering of others. The Hungarian-American psychologist Mihaly Csikszentmihalyi would presumably include his notion of 'flow' (also known as 'the zone') — in which a person performing an activity is fully immersed in a feeling of energised focus — as one of the vehicles of deliverance from wanting. When we are immersed in an activity, the actual goal of the activity fades into the background as the activity itself takes centre stage. Or, expressed in Schopenhauerian terms: the will is extinguished because it no longer has any purpose on which to focus. Anyone who has ever lost track of time while out walking, while painting, or even while gardening, will know this to be true.

So there are many different ways to extinguish the will and the suffering it causes. And I can add to the list one more — rather surprising — way, which Schopenhauer could never have imagined. And that is where this book comes full circle.

## The nirvana of locked-in syndrome

If there can ever be a description of the nature of the brain, then it must be this: the brain wants to achieve an effect, and the effect triggers the wanting. But what happens when the brain can no longer achieve any effects? When it no longer has access to the muscles and other organs of the body in order to set anything in motion — as is the case with locked-in patients?

For most people, the idea of being trapped in their own body is catastrophic, and can lead to nothing but despair. The reason for this is that we are creatures driven by will and so we cannot imagine the life of a locked-in patient from any other point of view. We see the totally paralysed patient lying in bed, being artificially ventilated, and feel only pity for that person and fear of ending up in the same position, because such patients are unable to do any of the things that people normally do. They cannot eat or drink, they cannot talk, they cannot even breathe unaided. If they want to raise their arm, or even just a finger, they cannot, because their body is as unresponsive as a corpse.

That sounds like something out of a horror film, and, indeed, a person who wants to do something but is suddenly unable to do it will undoubtedly suffer terribly at first. When a motorcyclist wakes up after an accident to find he is paralysed from the shoulders down, he will initially be aghast and horrified. However, the reason for this is that to him it seems that, only seconds before, he was in full possession of his ability to move and so he has had no time to become accustomed to his paralysed state.

He lives in *hope* (for which read: will) of regaining his ability to move.

But what about someone who has come to terms with their condition and abandoned any hope that it will change in the future? This is precisely the case with many locked-in patients: with time, they no longer rebel against their fate, becoming accustomed to and accepting their paralysis, sure in the knowledge that they will almost certainly never recover. In the truest sense of the word, they are hopeless, without hope, which Zen Buddhism teaches can lead to a life full of peace, joy, and compassion.

Laboratory experiments have shown that lower lifeforms cease any activity that no longer produces an effect.[19] No matter how much they are stimulated, they no longer respond. Their will to engage in that activity is extinguished even though, physiologically speaking, they would still have the energy available to perform the action. And this is no less true for the complex neuronal structure of the brain.

From an evolutionary point of view, this makes complete sense. Why expend energy and resources when there is nothing to be gained from it? Nature's rule is: investments must pay off — and if an effect can no longer be achieved, then it is better to save the energy and stop investing resources.

We recognised this tendency in our locked-in patients. We had to use various 'tricks' and technical wizardry to rouse them out of their state of inactivity so that we could communicate with them. But does that mean they were unhappy?

Although we found barely any sign of will in the brain activity of our patients, we also found no indications that they felt miserable or desperate. We saw nothing of the withdrawal typical of addicts or anyone with a passionate want for something. Their brains also did not resemble those of people with depression, who see no point to their actions, because they expect only negative effects, and suffer as a result of this. The reason for all this is that locked-in patients do not suffer, because they have lost both the possibility to engage in any action and the expectation of negative results.

When we managed to make contact with our locked-in patients, they turned out to be happier than the average, healthy population. However, that was at a point when they were finally able to communicate, and so such a metric should not be afforded too much significance and should not be assumed to apply to patients who remain cut off. However, it does seem to be a strong indication that locked-in patients, in their own special way, have arrived at that 'absence of will' proclaimed by Schopenhauer and by Zen Buddhists to be deliverance from the tribulations of life: free of any feeling of being driven and free from any feeling of withdrawal.

*What* really goes on inside these people is something we may never know. Perhaps they inhabit a dream world, akin to daydreaming during a long train journey, when the driver and the tracks take over all responsibility for targeted motion. Maybe their thoughts are similar to those of meditating yogis. It would certainly be interesting to compare the brain activity of someone immersed in meditation and someone with locked-in syndrome.

I have now moved into the realms of speculation, since we have not yet managed to prove that completely locked-in patients are in a state of will-less bliss. And perhaps that is a good thing, since otherwise, some people might want to enter this state deliberately.

However, our results, as described in this book, are enough to persuade us to become more cautious. They show that we should not make judgements about the happiness or otherwise of a person who is no longer capable of wanting, from our point of view as creatures driven by will. The call for such patients to be 'switched off' and for them to 'decide for themselves about life and death if it comes to it' and for legally binding living wills — though they were drawn up when the signatory still had the capacity to want — are all nothing more than proof of a lack of an ability to imagine ourselves in such a different situation. We cannot completely exclude the possibility that even completely locked-in and severely injured patients can still find happiness. There is more evidence indicating that this is the case than that it is not. Our brains are capable of anything — even doing nothing.

# Acknowledgements

Most of the research reported in this book, carried out by Dr Birbaumer and his team at his laboratory in Tübingen — insofar as it was published in specialist journals — was funded by the German Research Foundation (DFG). Our thanks go to the DFG and its highly competent specialist staff for that support, in particular Dr Anne Brüggemann, Dr Theodora Hogenkamp, Dr Manfred Niessen, and Dr Manfred Zimmermann.

# References

Birbaumer, Niels and Schmidt, Robert F., *Biologische Psychologie*,
    Heidelberg 2011.

Birbaumer, Niels et al., 'Learned Regulation of Brain Metabolism',
    *Cognitive Science*, 2013; 17(6).

Birbaumer, Niels et al., 'A Spelling Device for the Paralysed', *Nature*,
    1999; 398(3).

Carlson, Neil R., *Physiology of Behaviour*, Boston 2004.

Dworkin, Barry R., *Learning and Physiological Regulation*,
    Chicago 1993.

Gazzaniga, Michael (ed.), *The Cognitive Neurosciences*,
    Hong Kong 2008.

Hare, Robert D., *Without Conscience: the disturbing world of the
    psychopaths among us*, New York 1999.

Hohl, Ludwig, *Die Notizen oder Von der unvoreiligen Versöhnung*,
    Berlin 1984.

John, Erwin Roy, *Mechanisms of Memory*, New York 1967.

Schopenhauer, Arthur, *The World as Will and Representation*, (1819).

Slater, Lauren, *Opening Skinner's Box: great psychology experiments of the 20th century*, New York 2005.

Zeier, Hans (ed.), *Pawlow und die Folgen: Von der klassischen Konditionierung bis zur Verhaltenstherapie*, Reinbek, Germany 1987.

# Notes

1   Maguire, Eleanor A. et al., 'London Taxi Drivers and Bus Drivers: a structural MRI and neuropsychological analysis', *Hippocampus*, December 2006; 16(12), doi: 10.1002/hipo.20233.

2   Birbaumer, Niels and Schmidt, Robert F., *Biologische Psychologie*: p. 614 f.

3   Zeanah, Charles H. et al. The Bucharest Early Intervention Project: case study in the ethics of mental health research', *Journal of Nervous and Mental Disease*, March 2012: 200(3), doi: 10.1097/NMD.0b013e318247d275.

4   Narita, Kosuke et al., 'Relationship of Parental Bonding Styles with Grey Matter Volume of Dorsolateral Prefrontal Cortex in Young Adults', *Progress in Neuro-Psychopharmacology & Biological Psychiatry*, May 2010; 34(4), doi: 10.1016/j.pnpbp.2010.02.025.

5   Boyke, Janina et al. 'Training-induced Brain Structure Changes in the Elderly', *The Journal of Neuroscience*, July 2008; 28(28), doi: 10.1523/JNEUROSCI.0742-08.2008.

6    Bouchard, Thomas J., Jr, 'Genetic Influence on Human
     Intelligence', *Annals of Human Biology*, 2009; 36(5), doi:
     10.1080/03014460903103939.

7    Kotchoubey, Boris et al., 'Apallic Syndrome Is Not Apallic: is
     vegetative state vegetative?', *Neuropsychological Rehabilitation*,
     2005; 15(3/4), doi: 10.1080/09602010443000416.

8    Miller, Neal A., 'Learning of Visceral and Glandular
     Responses', *Science*, January 1969; 163(3866), doi: 10.1126/
     science.163.3866.434.

9    For more, see the University of Florida website: http://csea.
     phhp.ufl.edu/Media.html.

10   Ramos-Murguialday, Ander et al. 'Brain–Machine Interface in
     Chronic Stroke Rehabilitation', *Annals of Neurology*, July 2013;
     74(1), doi: 10.1002/ana.23879.

11   Muckli, Lars et al., 'Bilateral Visual Field Maps in a Patient
     with Only One Hemisphere', *Proceedings of the National
     Academy of Sciences*, July 2009; 106(31), doi: 10.1073/
     pnas.0809688106.

12   Feuillet, Lionel et al., 'Brain of a White-collar Worker',
     *The Lancet*, July 2007; 370(9583), doi: 10.1016/S0140-
     6736(07)61127-1.

13   Angell, Marcia, 'The Epidemic of Mental Illness: why', *The New
     York Review of Books*, 23 June 2011.

14   Sherwood, Chet C. et al. 'Ageing of the Cerebral Cortex
     Differs between Humans and Chimpanzees', *Proceedings of the
     National Academy of Sciences*, July 2011; 108(32), doi: 10.1073/
     pnas.1016709108.

15   Prüss, Harald et al., 'IgA NMDA Receptor Antibodies

Are Markers of Synaptic Immunity in Slow Cognitive Impairment', *Neurology*, April 2012; 78(22), doi: 10.1212/WNL.0b013e318258300d.

16   Simmons-Stern, Nicholas R. et al. 'Music as a Memory Enhancer in Patients with Alzheimer's Disease', *Neuropsychologia*, August 2010; 48(10), doi: 10.1016/j.neuropsychologia.2010.04.033.

17   Subramanian, Leena et al., 'Real-time Functional Magnetic Resonance Imaging Neurofeedback for Treatment of Parkinson's Disease', *Journal of Neuroscience*, November 2011; 31(45), doi: 10.1523/JNEUROSCI.3498-11.2011.

18   Slagter, Heleen A. et al, 'Mental Training Affects Distribution of Limited Brain Resources', *PLOS Biology*, May 2007; 5(6), doi: 10.1371/journal.pbio.0050138.

19   Koralek, Aaron C. et al, 'Corticostriatal Plasticity is Necessary for Learning Intentional Neuroprosthetic Skills', *Nature*, March 2012; 483(7389), doi: 10.1038/nature10845.